WEBSTERS PRACTICAL FORESTRY

WEBSTERS PRACTICAL FORESTRY

(A Popular Handbook on
The Rearing and Growth of Trees for Profit or Ornament)

A.D. WEBSTER

GREENWORLD PUBLISHERS

First Indian Reprint 2001

ISBN 81-87421-22-3

Published by
Greenworld Publishers
8/217, Indira Nagar
Lucknow - 226 016 (UP)
Tel. : 701433

Exclusively Distributed by
International Book Distributing Company
Chaman Studio Building, 2nd Floor,
Charbagh, Lucknow - 226 004 (India)
Tel. : Off. : 450004, 450007, 459058 Res. : 711172, 342829
Fax : 0522-458629
E-Mail : ibdco@sancharnet.in

Printed at :
Army Printing Press
33, Nehru Road, Sadar Cantt
Lucknow - 226 002
Tel. : 0522- 481164, 480546
E-mail : pressneversleeps@rediffmail.com

TO

MY WIFE

THE FIFTH EDITION OF "PRACTICAL FORESTRY"

IS DEDICATED

PREFACE TO FIFTH EDITION

To the Author at least, it is gratifying to know that four editions of *Practical Forestry* have been sold out, and that a fifth is urgently called for ; no other work on British Forestry having achieved a like popularity.

The whole has now been completely revised and several important chapters added, particularly with reference to Timber and the War, British Timber and Timber Trees, Afforesting Waste Lands, and The Education of Foresters, on both of which latter questions the Author had the honour of giving evidence before the Departmental Committee on Forestry of the Board of Agriculture.

A. D. WEBSTER.

REGENT'S PARK,

INTRODUCTION TO FIFTH EDITION

In my long experience of British Forestry, which, in a practical way, has extended over a period of forty years, I have become more and more convinced that in order to place it on a systematic and sound economic footing, State aid and the afforesting of large areas of comparatively waste lands are first necessities.

For the past thirty years I have not failed to urge upon the State, as well as private owners of suitable land, the pressing necessity of afforestation; and though in this matter a start has been made, yet this can only be looked upon as a faint, half-hearted attempt quite unworthy of our country and the vast interests at stake. As early as 1883, I drew attention to this matter in *Woods and Forests*, and at later periods in most of the leading journals and papers of the day; while in my evidence given before the Select Committee on Forestry, and in a paper contributed by special request to the Board of Agriculture, I went fully into the question, and pointed out what a boon to the unemployed, and how great a saving to the country would be effected by a well-organized scheme of tree planting.

Years ago I urged the Government to take up the question of the Larch disease, the ravages of which I then described as being little short of a national calamity. To the Highland and Agricultural Society of Scotland, and the Royal Scottish Arboricultural Society, I have contributed twenty-three papers on different topics connected with forestry, for which special medals have been awarded, while my *Practical Forestry* has now passed into a fifth edition.

In connection with the afforesting of waste lands, I have

travelled over the greater part of the United Kingdom and have examined much of the ground that could be set aside for this purpose, including the peat bogs of Ireland; while at altitudes up to 1,100 ft. I have formed plantations on the bare and wind-swept hillsides of Wales and Scotland, which to-day are proving not only a boon to the farmers in the way of the shelter they afford, but also a considerable source of profit to the owners.

The above investigations, combined with the examination of and reports on several of the largest woodland properties in this country, made at the request of the owners, have given me a wide insight into the forestry problem generally, but particularly with reference to our requirements in the near future, in view of the fact that the United Kingdom is by far the largest timber importing country in the world.

A. D. W.

CONTENTS

LIST OF ILLUSTRATIONS

WEBSTER'S PRACTICAL FORESTRY

CHAPTER I

FORESTRY AND THE WAR

DURING the war, and for long afterwards, there is bound to be a dearth of foreign timber, with the result that our home supplies will be taxed to the utmost. This, indeed, is already the case, and never before in the history of our country has the demand for British-grown timber been greater or prices higher than at the present time.

Plantations in England and Scotland are being cut down wholesale by the Government, while felling on private estates is going on with such rapidity that in a couple of years at most there is bound to be a dearth of all home-grown timber, and of pitwood in particular.

In the aggregate, our supplies are by no means great, the total area at present under woodlands being only a little over three million acres, to which, however, must be added the amount of field and hedgerow timber—a by no means inconsiderable quantity. Of Plantation timber much is, however, of very inferior quality and only suitable for rough fencing and mining purposes, and this applies generally to that grown throughout England, Wales and Ireland. In Scotland, however, there are large areas of coniferous woods that were planted for purely economic purposes which will yield a certain amount of valuable timber for pit-props and other immediate requirements. The felling of Scotch, Spruce, and coniferous trees will, however, be a blessing in disguise, for of late years both demand and price for this particular class of timber have been by no means great and have hardly warranted the owners of vacant land in undertaking the much-needed extension of our woodlands. Larch, on the other hand, is scarce and always in demand at a highly remunerative price, as also are Ash, Oak and Beech of large size and good quality.

It is useless to go into the question of how long our home supplies of timber will last, this depending largely on the volume of foreign importations and the duration of the war, but, judging from present demands, three years will find our plantations in a more or less depleted condition. The main question we have to consider and decide, without delay, is how our cut-over plantations are to be replaced by the replanting of exhausted woodlands and afforesting some of the waste lands of our country.

Continental experience has demonstrated that, from a commercial point of view at least, State-owned forests are preferable to such as are owned either by public bodies or by private individuals. The resources and continuity of a nation will always make the State the best custodian of forest property ; indeed, only the State can acquire the necessary land on the most favourable terms, and in sufficient quantity for the purposes of extensive afforestation. Private individuals—or, indeed, public bodies—labour under many disadvantages in this respect, not the least being the long period required—in most cases from fifty to sixty years—before the money expended in planting can be even partially recovered.

Contrary to the conditions obtaining in the raising of agricultural crops, long periods have to elapse before the forestry harvest can be reaped. It will be obvious, therefore, that extensive tree-planting is quite beyond the power of the private individual unassisted. It is a State business, in which systematic methods of cultivation, and large wooded areas are first necessities ; and, unfortunately, in this country commercial forestry is but little understood—in fact, it may be described as an unknown industry.

Extensive plantations of from a thousand to several thousand acres each, in compact blocks, are required, for it is only in dealing with such areas that the planting, tending, thinning, and conversion of the timber can be most economically and profitably carried out. From the purely economic point of view, the many small plantations dotted over the face of our country are worse than useless, though exceptions might be noted, especially in Scotland, where a few woods are planted and managed on a commercial basis.

With compact blocks of forest 2,000 to 4,000 acres in extent, and with timber crops of mainly the same species in each, a continuity of supplies could be guaranteed, which under existing circumstances is quite out of the question. In many outlying districts all over the country that are far removed from road and rail it is difficult—in some cases impossible—to dispose of the usually small amount of timber that is periodically cut down; but were large quantities of the same kind and a continuity of supply ensured, merchants would be tempted to make special transit and other arrangements, as well as to offer a remunerative price for the timber, while railway companies would no doubt provide cheaper facilities for its transport.

More than once I have been asked by owners of woodlands to recommend buyers of good Larch, Ash, and other timber, the demand for which far exceeds the supply; but after negotiating have invariably been told by the merchants that the quantity offered was too small to allow of special facilities for delivery being provided, the timber being far removed from road and rail, but that if a specified number of trees could be guaranteed annually for a number of years they were quite prepared to buy. Such cases occurred in the South and West of Ireland, and in remote parts of Wales and Scotland. These, then, are cases in which a continuity of supply—such as would be quite possible if an extensive scheme of afforesting was carried out—would ensure speedy sales at fair rates in places where at present it is difficult, if not impossible, to dispose of the small quantities of timber, even at ruinously low prices.

The question, then, before the nation is: How, and by whom, is afforestation to be carried out on a scale commensurate with our necessities as by far the largest timber-importing country in the world?—larger, indeed, than all the countries of Europe put together. In answer, and without the slightest hesitation, I would say that the State should acquire and plant suitable lands at the rate of 40,000 acres annually for a period of twenty-five years, or 1,000,000 in all.

When we consider that the total area of woodlands in this country is only a little in excess of 3,071,000 acres, that fully 17,000,000 acres of waste lands exist, and that we annually import over 10,000,000 tons of timber at a cost of about £25,000,000, the necessity for an increased area of woodlands, so that a portion at least of this vast sum may be kept at home, will be apparent to all, the more so as a dearth of timber is imminent and outside supplies are being rigidly conserved, while our home demands are ever on the increase.

In order to carry out my proposed scheme of planting a million acres during the next five years, at the rate of 200,000 acres annually, profitable and healthy employment would at once be found for several thousand workmen. The question of transporting, housing and otherwise dealing with these workmen has been brought forward as the most serious drawback to the scheme, but, personally, having had to deal with such cases I can see no insuperable difficulty in the undertaking. Surely, if our railway and water companies, as also private landowners, can deal with hundreds of men in remote mountain districts that are far removed from road and rail, the Government could make the necessary arrangements for the various bodies of workmen that would be employed for afforesting purposes. Nor must tree planting be considered as a new departure for unemployed labour, as in the formation of a large plantation on a dreary exposed hillside in Wales, the whole of the work, including clearing the ground of rough surface growth, draining, pitting and planting, was most successfully carried out by detachments of the unemployed.

The general physique of Army and Navy men and the discipline and hard work to which they have been subjected during the war will render them peculiarly suitable for carrying out the various operations connected with the formation of plantations. Convalescent soldiers and sailors could also find healthy employment in the various lighter tasks which go hand in hand with afforestation, such as clearing the ground of rough growing vegetation and lifting and distributing the young trees.

Several suggestions have been made for providing employment for our returned soldiers and sailors, but much of this proposed work is, so to speak, invented for the purpose, and would probably never be seriously considered except for the exigencies of the case. Now I am quite of opinion that afforesting waste lands offers a sensible system of employment, for it is now generally admitted that a largely increased area of our woodlands is an imperative and pressing necessity, and what is of equal importance, the undertaking, if wisely entered upon, would not only increase the value of such lands fourfold but form the nucleus of an ever-increasing revenue of the State.

But this is not all, for apart altogether from the question of immediate labour, what an industry would be opened up in years to come by the planting of waste grounds! In the first instance there would arise the necessity for clearing, fencing, draining, and planting the ground, subsequently the tending of the plantations in various ways would give employment. Thinning would commence about the tenth year, after which the erection of sawmills and the conversion of the timber would open a vast and ever-increasing industry, providing highly remunerative work to thousands of the unemployed.

There are other ways in connection with forestry in which the unemployed could be usefully and profitably set to work, such as in preparing osier beds, planting dogwood for gunpowder charcoal, and in the formation and stocking of tree nurseries, the produce of which could with advantage be used in the formation of plantations. Osiers for basket-making, charcoal for gunpowder and heating purposes, and seedling plants for afforestation will all be greatly in demand after the war when foreign supplies will not be forthcoming.

Regarding the most desirable centres at which to commence planting operations, I would suggest those counties where the greatest areas of waste lands exist, and where, in addition, advantages are offered in the matter of cheap land purchase, as well as in a demand for and easy removal of the produce. Thus we have :—

England . Yorkshire and Northumberland with 1,010,924 acres.
Scotland . Inverness and Argyleshire with 3,087,312 acres.
Wales . Breconshire and Merionethshire with 461,320 acres.
Ireland . Donegal and Kerry with 657,337 acres (exclusive of 172,436 acres of bogland).

From these figures it will be seen that in case of necessity we could get all the ground required for this afforesting scheme in two of the counties of England and Scotland, or in three of Wales and Ireland.

The cost of procuring suitable land for afforesting purposes need not be considered as any obstacle to the scheme. From extensive inquiries made the price on an average would not be greater than £2 per acre. Through the kindness of the agent on the Gwydyr Estate, in Carnarvonshire, I have been allowed to look over the sale contracts on several of the properties, and from these I find that 7,412 acres were disposed of at an average price of £2 2*s.* 3*d.* per acre. The ground was excellent for the production of timber, as the Larch on adjoining lands clearly evidenced. Again, the Crown recently purchased 12,500 acres in Scotland at the modest rental of £2 per acre. Many other instances could be quoted, but the above suffice to show that land in every way suitable for the production of high-grade timber can be bought at probably less than £2 per acre.

The cost of forming plantations has been very carefully

considered, and for all practical purposes may be put down at £5 per acre, taking the British Isles as a whole.

Regarding financial returns from tree planting, there is overwhelming proof that land worth only from 1*s.* to 3*s.* 6*d.* per acre has been made to realise as much as 20*s.* per acre for fifty or sixty years, with a final crop worth from £50 to £75 per acre.

Taking all the above points into consideration—the price of land, the cost of planting and the financial returns—it will amply repay the State to plant up uncultivated and waste lands. There should be no loss of time in setting about this work if we are to provide suitable employment for our returned soldiers and sailors and avoid the threatened timber famine which, according to well-informed quarters, is fast approaching.

CHAPTER II

COLLECTING AND STORING TREE SEEDS

OWING to the war there is little doubt that for several years to come adequate supplies of seeds and young forest trees will not be available from their usual sources on the Continent. Vast quantities of both are annually sent to this country from Germany alone, while from other parts of the Continent coniferous seeds in particular are largely imported—in fact, we are almost wholly dependent for both seeds and seedlings on foreign supplies.

Under these circumstances, it behoves the owners of woodlands in every part of the country to collect seeds, particularly of such trees as our plantations are mainly composed of. After these have been properly harvested, they should be stored away in suitable places till seed-sowing time in the spring. To some extent this will be a new forest industry, though for long, the seeds both of Scotch Pine and various hardwooded trees, particularly the Oak, Ash and Elm, have been annually collected on various estates throughout the country, and the young plants raised from such stock have given every satisfaction when planted out permanently.

For several reasons, however, we cannot compete with foreigners either in the production of seeds or in raising young forest stock. For some years to come, owing to the uncertainty of supplies from abroad, both seeds and seedlings are bound to be scarce and expensive.

In the past, far too little attention has been bestowed on the collecting and harvesting of the seeds of trees and shrubs, the result, in not a few cases, being weak and unhealthy plants and an uneven and irregular crop. The best seeds, it should be remembered, are those collected from healthy trees in the prime of life, and grown under conditions favourable to their perfect development. An unhealthy tree will often bear a heavy crop of seed, but although the inducements to collect such are great, they should be discarded, those from the most robust specimens in the prime of life being chosen in preference.

Regarding the best way of collecting tree seeds little need be said, the exigencies of the case determining the best method to be adopted. The seeds of not a few trees may be collected as they fall, and this is especially the case with those of the Oak, Beech, Elm, etc., all of which may be swept into heaps and gathered in quantity from beneath desirable trees.

In the case of the various Coniferæ this method of seed collecting will not answer—indeed, in the majority of instances, the seed should be gathered, or rather picked, from the trees just before they became fully ripe, as in falling they get loose from the cone-scales and are lost. When collecting the cones of coniferous trees, a long, light hooked staff with which to draw the branches towards one can conveniently be used to procure an abundant supply. A bag or satchel should also be in possession of the seed collector, into which may be put such kinds of cones as fall readily apart, as the seeds from these are easily lost. Sometimes, as in the case of rare seeds, and when only a few cones are borne near the top of the tree, the seed collector must have recourse to climbing; but, in such cases, in order to avoid injury to the bark, he should be provided with a pair of elastic shoes or slippers. Great care is required in the collection of such seeds as those of *Abies nobilis* and *A. nordmanniana*, the cones, when fully ripe, falling to pieces on the slightest touch. This, however, applies equally to almost every species of Abies, whereas, with the Pines and Spruces, the cones remain intact for an almost indefinite period of time, even though the seeds may have fallen out on becoming ripe.

The proper harvesting of tree seeds rarely, except in the case of experienced nurserymen, receives sufficient attention, although this operation should be as carefully attended to as in the case of the seeds of any other form of crop. After being collected, the seeds of all trees, unless such as are mixed with sand for the purpose of rotting, should be thinly and evenly spread out in a sunny spot, until thoroughly dry. They may then be deposited in a cool, airy place, and in thin layers, until wanted for sowing. An occasional turning is all-important and should never be neglected. The smaller and less common seeds may, for convenience sake, be hung up in calico bags, but they, too, should be occasionally examined to prevent dampness and heating.

The number of plants of various kinds that may be expected from a bushel of seed of average quality varies very much, and may be approximately given as follows: Horse Chestnut, 2,500; Oak, 6,000 to 8,000; Spanish Chestnut, about 3,000; Walnut, 5,000; Norway Maple, 12,000; Sycamore, about 12,000; Ash, 14,000; Beech, 10,000; Elm, 1,000; Birch, fully 16,000; Holly, 17,000; Scotch Fir, 9,000. To 1 lb. of seed: Spruce Fir, about 9,000; Larch, 3,000; and the Cluster Pine, Silver Fir, and some others, from about 500 upwards.

For convenience in regulating orders for sowing, the following table will show at a glance the approximate and relative number of seeds of the various commonly cultivated forest trees contained in 1 lb. weight:—

Abies nobilis	**about**	**19,400**
Abies nordmanniana	,,	**10,000**
Ash	,,	**6,800**
Beech	,,	**2,700**
Douglas Fir	,,	**95,200**
Horse Chestnut	,,	**36**
Hornbeam	,,	**9,968**
Larch	,,	**65,000**
Lawson's Cypress	,,	**131,400**
Lebanon Cedar	,,	**10,800**
Norway Maple	,,	**4,600**
Oak	,,	**100**
Pinus Austriaca	,,	**35,000**
Pinus Laricio	,,	**43,000**
Pinus Pinaster	,,	**12,000**

Pinus Sylvestris	about	75,000
Spruce	,,	64,500
Silver Fir	,,	14,960
Sycamore	,,	4,624
Walnut	,,	36

These figures must only be taken as approximate, the seed of various trees of the same species seeming to vary in the number to the pound in a marked degree. The results of careful analysis of one or two kinds may be cited as examples. In one case the number of seeds in 1 lb. weight of Scotch Fir was 69,600, while in another it had increased to 90,600 ; and in the Larch the numbers were 33,900 and 68,000. These differences are, however, mainly due to the individual seeds being weightier in one case than in another, probably owing to the age and health of the tree from which they were collected, the situation and exposure to which it was subjected, etc. However, for all nursery purposes the above figures may be accepted as a fair standard.

The time of collecting and the after-management of the different forest seeds vary so much that a brief description of those kinds most commonly planted will be found useful :—

Alder seed should be gathered from the trees in October and sown in spring, say May.

Ash seeds are ripe in October, when they should be collected and kept in moist sand during the winter, and sown in March.

Austrian, **Corsican** and **Weymouth Pine** seeds are treated in every respect like those of Scotch Fir, varying the kiln heat according to the looseness of the cone-bracts.

Beech seeds are collected in October and November, placed in sand, and sown in April. The young plants are readily affected by frost, and the seed should, therefore, not be sown earlier than the time mentioned.

Birch seed must be collected from the trees just before it becomes ripe in August, otherwise it is scattered broadcast and lost for cultivation. March is the time for sowing.

Cupressus Lawsoniana seed is usually ready for

collecting in October, but should not be sown before the first week in April.

Douglas Fir seed is, in most cases, readily removed from well-ripened cones by threshing or by pulling the cone to pieces, but, in some instances, particularly where the quantity is large, kiln-drying is resorted to. The seeds are ripe in December and should be gently watered and sown in May.

Elm seeds are ripe in June, when they may either be sown at once, or dried and kept in stock for planting in March and April.

Hawthorn seed, or berries, may be sown when collected, or the outer coating rotted off by keeping them during the winter in moist sand.

Hazel nuts may be collected in autumn and sown at once, or kept till spring.

Holly berries require to be placed in sand for about eighteen months so as to rot off the fleshy outer coating, and may be sown in March. The mixture of sand and berries, which should be about in equal proportions, must be turned frequently. They are usually sown with the sand in which they have been lying.

Horse and **Spanish Chestnut** seeds may be taken together, the method of collecting and sowing being very similar in both cases. They are ripe by the middle of October, and may either be sown at once or kept till spring. One seed to 4 square in. will be close enough.

Larch cones, when ripe, are of a rather bright brown colour and require to be collected from the trees. This should not, however, be done till spring, though occasionally they are gathered in December. They part with the seeds far more readily than those of the Scotch Fir, and consequently require less heat when in the kiln.

Maple seeds are ready for collecting about October, and should not be sown till the beginning of April.

Mountain Ash, indeed, all the Pyrus family and others of a like kind, require the berries to be placed in sand, and when the outer fleshy coating has rotted away they may be sown either in autumn or spring.

Oak.—The acorns may be gathered or swept from the ground in November, and either sown at once or stored away in a cool, dry place till Spring. One acorn to every 4 square in. will be ample in the seed bed. Sow in spring or autumn.

Scotch Fir cones are better not collected till early in January, and the time may even be extended till March. When quite ripe they have changed from bluish-green to a light, grey colour. As the cones part tardily with the seeds, artificial means have to be resorted to, the cones being placed thinly over a kiln heated to a temperature of from 75° to 112°. They should be turned every third hour, and after about thirty hours the kiln should be cooled down and the cones extracted as quickly as possible. By beating with a flail the seeds are readily removed from the cones, but it is best to do this before the cones have cooled down or immediately they are removed from the kiln. The seeds are then swept together and collected, and stored away until wanted for sowing. When not required for sowing at once, the seeds should be thinly spread out on the floor and slightly moistened with water from a fine rose watering-can. They should then be turned about until perfectly dry before being stored away.

Silver Fir seed does not require much, if any, artificial heat to cause it to part from the cone. By placing the cones in the sunshine, and heating and turning freely, the seeds come out without much trouble. In all cases, however, wherever possible, it is wise policy to dispense with artificial heat or kiln-drying as, unless this is carried out most carefully, the vitality of the seeds is greatly impaired thereby.

Sycamore seeds are ready for gathering in October, but should not be sown till the end of March or beginning of April.

Walnuts are collected, when ripe, in autumn, and sown in late spring.

Yew seeds are usually washed of the pulpy matter before being sown.

In the case of large seeds, such as those of **Araucaria**

imbricata, Pinus Sabiniana, and **P. macrocarpa,** the best way is to cut the cones to pieces and carefully remove the seeds. This operation should be performed with great care, so that the hard seed coating may not be injured.

With conifers in general I have invariably found it the best plan to allow the seed to remain in the cones until wanted for sowing. By keeping the cones in a cool, dry place, and occasionally turning them over, there need be little fear but that the seeds will turn out well.

CHAPTER III

PROPAGATING TREES AND SHRUBS

THERE are several methods of propagating trees and shrubs: such as by seed-sowing, from cuttings or layers, and by budding and grafting.

From Seed.—This natural process of reproduction is that most commonly adopted where large numbers of trees are required—indeed, certain species, particularly of the Coniferæ, cannot successfully be raised in any other way. The preparation of the seed-beds is a point that deserves far more attention than it usually receives, and that not only on economic grounds, but in view of the general appearance of the nursery borders as well. To tumble the seeds indiscriminately into the ground as if they had fallen in showers from the trees is highly objectionable; and just as censurable is the too-oft-repeated practice of sowing these in rough, cloddy, and ill-prepared ground.

Systematic arrangement in the laying-out of the beds, as to the number and requirements of the future seedlings, should also be attended to.

The ground intended for seed-beds should be trenched or deeply dug up; but this operation would, for the mellowing and cleaning of the soil, be better performed the autumn before than at the time of sowing down.

In any case, just before sowing, the ground should be carefully turned over, all hard clods being broken down, and large stones raked off, the surface soil to the depth of 3 in. or 4 in. being made as fine as possible. Dry weather must be chosen for the formation of seed-beds, as also, indeed, for the sowing of the seeds. The beds are marked off and prepared as follows:—A light, strong line is stretched along the ground at, say, 3 ft. from the boundary fence or path, and, after being fixed at each end, the first alley, or path, is marked out by treading the soil alongside of the line. The length of the bed thus marked off on one side is quite immaterial, but the width is of great importance, and should never exceed 4 ft.; 3 ft. 6 in. is nearer the mark.

Adjoining this first line, a bed the above width is marked off with a peg or stake at each end, the line reset and a second alley marked off. The alley, or path, between each couple of beds need not exceed 15 in. in width, this being for the sake of convenience in sowing, weeding, watering, shading, and otherwise attending to the bed, and its occupants. Outside this follows a second bed, and so on, until the desired number has been formed. Some persons raise the seed-beds a little above the level of the paths, but, except in very retentive or damp soils, this is not to be recommended. The beds should in no instance, however, be below the level of the paths. There are two methods employed in opening up and preparing the beds for the reception of the seeds either of which, if carefully gone about, is well suited for the exigencies of the case. My own way has always been to level and smooth the surface of the bed with a small-toothed rake, and after sowing the seeds to cover lightly with fine soil and ashes carefully sifted over the bed by means of a small-meshed riddle. This plan has many advantages, not the least important being the covering of the seeds to an equal depth, and the employing of only the finest class of soil. The other method is by using what is termed a "cuffing-board"—that is, a board about 8 in. wide, placed on a handle, which is inserted in the centre, towards the back, the handle being fully 5 ft. long. A skilled person is required to use this tool, who stands in the alley, first on one side of the bed, and then on the other, pushing or drawing towards him from the surface of the bed a thin coating of soil along its full length and from about two-thirds of its surface.

As to the amount or depth of soil taken off, this is regulated by the particular kind of seed to be sown, as well as the nature of the soil.

The next matter, the depth at which the seed should be sown, is one of great importance. Usually seeds are sown unnecessarily deep. In the majority of cases a safe guide is to place the seed, of whatever kind it may be, about three diameters below the surface of the soil. The conditions most favourable to germination are moderate dampness, abundance of air, and a temperature of about 45°. In order to ensure these conditions, the depth at which the seeds are placed will be seen to be of the greatest moment. Generally speaking, the less seeds are covered, consistently with their receiving a sufficient supply of moisture, the better, and but for their destruction by birds, many of the smaller seeds would vegetate just as well if cast upon the ground-surface, and pressed in, as by being covered with soil. The following interesting experiments with Scotch fir seeds have been made :—

> "Those buried one-fifth of an inch came up first, but were subsequently less vigorous—they soon, however, acquired vigour ;
>
> "Those covered from one-third to one-half of an inch came up more slowly, but evenly and strong ;
>
> "Those buried from two-thirds of an inch to one inch came up in deficient numbers ; and
>
> "Those buried from one to two inches never showed any signs of germination."

These statements are equally applicable to the sowing of seeds in general but especially if their relative size and the hardness of their covering are taken into account.

Immediately after the bed is prepared the seed should be sown, the amount used varying according to quality, which latter may readily be tested by examining the embryos of a dozen seeds picked up at random from the heap. To ascertain whether seeds are good, the simplest way is to cut the sheath open with a sharp penknife, when the kernel ought to completely fill the entire coating or shell. Small seeds might be crushed by the nail, and if good, will leave traces of moisture or emit an odour of turpentine. Another method is to place, one after the other, say a dozen seeds taken at haphazard from the heap, on a red-hot iron. If good they will turn about with a cracking report; but if otherwise, combustion is slow and smoke is given off.

Seed may, however, be considered good if the germinating test gives, say, 70 per cent.

The quantity of seed to be sown on a given area will depend mainly on the quality and particular species.

When sown broadcast, about a pound weight of larch, and full half a pound of Scotch, is allowed per 100 square ft. of seed-bed.

Generally, however, drill sowing is resorted to, and in this case the following may be considered as about an average of the quantity used per 100 square ft.:

Alder	7¾ ozs.
Ash	5 ,,
Austrian Pine	4¼ ,,
Birch	6½ ,,
Elm	3½ ,,
Hornbeam	2¾ ,,
Larch	7½ ,,
Maple	5 ,,
Scotch fir	2¾ ,,
Silver ,,	13 ,,
Spruce ,,	3½ ,,

Sometimes a small quantity of guano or other manure is sprinkled over the seeds, after which they are gently pressed down with a light roller, and the soil, which was temporarily deposited along the margin of the bed, replaced by the "cuffing-board" or riddle. When seeds are sown early, and with care, on well-prepared ground, watering is seldom necessary; but still, with small seeds and in very dry ground, an occasional sprinkling the last thing at night is highly beneficial. Great care is, however, necessary to ensure the soil being thoroughly moistened. Protection from mice and birds must be afforded in some such manner as by coating the seeds with red lead; and an occasional shading of the bed in very warm weather might be advocated. There are not a few kinds of seeds, such as those with a hard covering, that it is almost imperative to steep in water before sowing. Larch seed, in particular, can be soaked for a week without fear or harm, and many leguminous seeds for two or three days with great advantage. Some seeds do not come up until the second year, such as the ash, pyrus, thorn, etc., and in these cases it is but a waste of ground to sow them as collected. They should be mixed with sand, as before directed, and stored away for a year before being sown.

It should also be remembered that, as a rule, the sooner seeds are sown after being collected, the stronger will be the young plants.

Soon after germination, hand weeding must be commenced, and this may most successfully be performed after a shower of rain, as the young plants then suffer less from loosening of the soil, consequent on pulling out the weeds, than would be the case in dry and warm weather.

The above method of raising plants from seeds will be found suitable in most cases, but in respect of choice or half-hardy trees the shelter afforded by a frame is usually provided. In this case the seeds are sown in pots or shallow boxes, according to quantity, and placed in an unheated frame.

The best time for seed-sowing is from the first to the fifteenth of April.

Transplanting the young seedlings may be taken in hand usually after the first year, but no hard-and-fast line can be laid down, so much depending on the season and rate of growth of the young plants. The experienced eye can always tell when seedlings should be lined out. The best time for this operation is after the spring frosts are at an end, for, if planted out in the autumn, the frost is apt to lift the tiny seedlings wholesale from the ground.

After this they should be transplanted at least every second year until planted out permanently.

By Cuttings.—Propagating trees and shrubs from cuttings is at once a simple and inexpensive way of getting up a stock of such kinds as may be increased in that way. The best months are August and September, at which time the temperature of the earth and air are equal, and roots are most readily emitted.

Choose a sheltered and partially shady border for the

insertion of the cuttings, and be sure that the soil is not stiff and water-logged, but fine, rather inclined to sandy, and moderately dry. It should be well forked over and cleaned of weeds, large stones being also removed.

The cuttings may be made of the present year's growth, with a small portion—" heel " it is generally termed—of the old wood, and cut clean across beneath where a bud or shoot has protruded. Until a sufficient quantity has been made, they should be stuck loosely in a heap of sand or soil, and in the shade. Cuttings should be taken from the sunny side of a tree or shrub.

In planting the cuttings, stretch a garden line along the previously prepared ground, and, having removed a trench about 6 in. deep, by cutting in a perpendicular manner with a spade along the front of the line, proceed to place the cuttings not too closely together, and leaning somewhat backwards in the trench cut out. This being done, it is always well, so as to hasten the formation of roots, to sprinkle a small quantity of sand or grit along at the base of the cuttings, at the same time regulating these as to their distances apart and upward inclination, the remaining portion of the trench being filled up with the soil formerly removed. Firm tramping is one of the chief points to be observed, as cuttings will not take at all quickly in loose soil. The line should then be moved forwards about 12 in. and the same method of procedure followed until the whole of the cuttings are inserted. For such shrubs as the common and Portugal laurel, privet, box, euonymus, laurestinus, aucuba, etc., the above method will be found the quickest and best for raising young plants from cuttings. It should be stated that cuttings taken from the sunny side of a tree or shrub always root more freely, and turn out a greater number of plants, than such as have been cut from the shady side—indeed, this point should be carefully borne in mind and acted upon. Conifers, such as the various species of Cupressus, Juniperus, the Wellingtonia, etc., are also readily raised from cuttings inserted in the ground out of doors. It is, however, well to put such cuttings in boxes, so that they have the advantage of being placed in an unheated frame, this greatly enhancing the speedy formation of roots. Two or three large frames are always useful in the nursery for the reception of choice cuttings, seeds, etc., and if boxes, say, 24 in. by 14 in. by 5 in., are made just to suit the dimensions of the frames, a great number of young plants, cuttings, and seeds of the less common kinds will find a beneficial protection. The boxes should be made of stout wood, and have six holes bored in the bottom of each, so that the excess water may readily pass away. Before planting these, fill each with a mixture of half sand, half soil, broken finely down, the cuttings being inserted in small holes made by a sharp-pointed stick. A thin coating of sand placed on the surface will greatly facilitate the process of rooting. Cuttings of coniferous trees should have 2 in. of ripened wood attached to them, and be inserted 4 in. deep in the ground; while, for most others, the cuttings may be 8 in. long, 3 in. being inserted in the soil.

From Layers.—For increasing game coverts, shrubberies, or some of those trees and shrubs that cannot well and speedily be raised from cuttings or seed, layering will be found a convenient and cheap method. There are several other advantages too, as no protection of any kind is necessary, and the risk of loss is reduced to a minimum. It is, however, generally resorted to as a means of increasing such plants as cannot readily be raised from cuttings, and applies with greatest force to the rhododendrons and azaleas, the magnolias, coniferæ of various kinds, and several species of our hardy deciduous trees.

The operation of layering is very simple, and consists merely in bending down the outer branch of a tree or shrub and bringing it in contact with the soil, it being there held in position by means of a hooked peg. Before fastening the layer with the peg, it is well, however, to remove a portion of the bark from that part of the layer that is to be brought in contact with the ground, this arresting the flow of sap and hastening the formation of roots. The soil beneath the tree or shrub should be first loosened, and, if found hard and stiff, a small quantity of sand and leaf soil might with advantage be added. The same principle is carried out in all kinds of layering, but the position of the plant will alter the method of bending over and bringing in contact with the soil. In the case of layering coniferæ it is not only necessary to bend over and peg down the branch, but, so as to form a leader and assist in gaining an upright position, it should also be tied or staked in a vertical position. Generally speaking, trees and shrubs require two years before the branches that are layered are sufficiently well rooted to be removed from the parent, but a few kinds require nearly double that time before they can be severed with safety. A good plan is, after the second or third year, to cut through the layered branch behind the point that is brought in contact with the ground, and to let it remain in position for another year. It may then, except in a very few cases, be lifted and lined out with others in the nursery border, the object being to get it into a good habit of growth before being finally planted out.

Most varieties of trees and shrubs can readily be increased by layering the side branches in the way above described, and it is a good plan to cover over with a spadeful of earth that part of the branch brought in contact with the ground.

By Grafting.—This may best be described as a makeshift method of increasing trees and shrubs. It has only a few advantages, and many disadvantages, as is well known to every one on a large estate who goes in for collections of the less common kinds of trees and shrubs.

If a plant cannot be increased conveniently by any of the above methods grafting may be resorted to, but the work should only be entrusted to those who have great practical knowledge of the art, and who have paid particular attention to the evils attending injudicious grafting. Some of the errors in grafting may be cited as follows: Working on too tall stocks whereby an unnatural appearance and continual source of danger in windy weather are brought about; grafting small scions on large stocks, and grafting evergreen on deciduous species—all of which have tended, in a marked degree, to lower the art in the opinion of horticulturists generally.

The grafting of conifers may be performed at one of two seasons: either early in spring or about the month of August. Two-year-old stock are mostly to be preferred, so that the stock and scion may be of as near a size as possible. Several methods of grafting are adopted, but that generally practised is what is known as "side grafting." This consists in making a clean cut in the stem, downwards, for about an inch in length. A transverse cut is made at the termination of this first cut, whereby a piece of the bark wood is removed from the side of the stock. The scion is next prepared by being cut off square at the end, and one side of the wood shaved off for about an equal length to the cut on the stock. Place the scion in the notch made on the stock, and be careful that the barks, on one side at least, come neatly together, for on this hangs the whole fate of successful grafting. The scion should, indeed, be made to fit as nearly as possible and replace the wedge cut from the stock. Tie firmly and carefully with bast matting, and cover over with prepared clay or grafting wax. Another method consists in making an incision obliquely in the stock, the scion being so prepared that one side of it forms a sharp edge, in order that it may be readily inserted therein, the bark of the scion and stock coming neatly together. Tie with matting as before directed, and cover over with clay or grafting wax. Inarching, or grafting by approach, consists in bringing the scion into union with the stock without detaching it from its own stem, and the separation is not made until the new connection has been formed.

Budding.—This is usually performed in July, and in the case of certain shrubs has been found useful. The bud should, in addition to its strip of bark, have a portion of the cambium attached, as unless this be present to unite with the cambium of the stock the operation will fail.

The bark of the stock is cut into a T shape, opened with the end of the budding knife, and the bud slipped neatly in. Bast matting will serve for tying, until a junction takes place, which is usually effected in five or six weeks.

CHAPTER IV

THE HOME NURSERY: ITS FORMATION AND MANAGEMENT

In the course of my experience on large estates where planting has been extensively engaged in, few things have been more particularly impressed upon me than the importance of having a well-managed and well-stocked home nursery, where the propagating and raising of trees and shrubs required for forest and ornamental planting may be taken in hand. The numerous advantages of a home nursery, especially where planting is extensively carried out, are now so well known and appreciated by every proprietor of large estates that comment on this subject seems almost unnecessary.

Where ornamental planting, the formation of woods and plantations, game coverts, or hedging is performed on an extensive scale, the convenience of a home nursery cannot be over-valued, the plants being at hand when wanted, of the size and in the quantity required, and known to be well rooted, sturdy, and free from disease.

The advantages in these cases are too well known to require comment, and plants, more especially those of a large size, sent from even a short distance by either road or rail, cannot be expected to succeed equally with those raised and planted on the same day. The extra soil, or ball, with which large plants can be removed for a short distance is also much in their favour, but it is next to impossible to retain this where packing and transit have to be resorted to.

It is well known that too sudden a change from rich, well-sheltered nursery borders to bare, exposed hillsides often proves fatal to young plants; and, when we consider that few public nurseries are at a greater elevation than about 500 ft., the necessity of proprietors rearing their own stock, whose plantations are, perhaps, upwards of 1,000 ft. above sea-level, will the more readily be seen. There are certain difficulties to contend with in planting high-lying ground, more especially if the soil is poor and thin, and the situation exposed, and in these cases the advantage of using hardy plants that have frequently been transplanted in a well-chosen home nursery at a fair altitude is very noticeable, especially when contrasted with others that have been grown under more favourable circumstances and in a sheltered position. Some plants seem better adapted than others for this removal, but in the majority of cases the shock sustained by transferring from low-lying ground to that at a great elevation is only too apparent, and one from which the plants seldom recover. The same holds good in the case of seaside and town planting.

Of late years in particular, a good deal of comment has taken place as to the necessity of rearing trees from seed sown on the site of the future plantation, and although the suggestion has many points in its favour, still artificial planting is better adapted to the wants of our country, and is not at all likely to be superseded by natural reproduction, which is more fitted for countries differently situated from our own.

The nursery treatment of plants is, therefore, sure to remain a prominent feature of British forestry, and this being the case, the soil and situation, as well as the most successful treatment of these, with a view to producing plants suitable for the positions they are intended to occupy, will require due consideration. This will vary much according to the situation of the estate and ground to be planted. In choosing the site of a home nursery, a great deal will depend on the general elevation and exposure of the estate. The situation should neither be too much exposed nor yet too sheltered, and should have a southern or western aspect; for, although too sudden a change from sheltered to exposed ground often proves fatal to young trees, this should not altogether form a criterion for rearing them in situations unfavourable to the development of strong, healthy plants. The soil should be good, friable loam, on an open, porous subsoil; but the quality of ground required for different seedlings is so diversified that it is next to impossible to suit all within the small bounds required for a home nursery.

As water is indispensable where seedlings are raised, as well as for numerous other purposes in the nursery, it is well to have provision made for a continuous supply, either by a stream running through the ground, or in close contiguity to it, or by having a pipe laid on from the main water-supply.

From six acres to ten or even fifteen acres will be found sufficient nursery ground for most estates, but it is always advisable to add a little more than is really required, so that the brakes may not be all under forest trees at the same time, but undergo, when found necessary, a course of green crops, which will not only enrich, but clean, the ground and leave it in good condition for replanting with seedling forest plants, bearing in mind that farmyard manure should always be applied first to the green crop, and never directly to the plants themselves. When a plot has become impoverished by repeated croppings of forest trees, a heavy coating of well-decomposed farmyard manure should be applied, and the ground planted with potatoes, or sown down with turnips. This has an almost magical effect in improving, regenerating, and cleaning the ground, and leaving it in the best possible condition for receiving a crop of forest plants. Land intended for nursery ground should be thoroughly trenched to the full depth of the soil, taking care, at the same time, that the best soil is kept within a reasonable distance of the surface, and, where necessary, heavily manured or enriched by the addition of lime, vegetable soil, or loam as the case may be.

In laying out the ground into brakes it will be found convenient to have these either square or rectangular in shape and, if possible, parallel with each other. The brakes should be of different sizes, and divided from each other by walks or hedges, but the fewer of the latter the better. It is well for convenience sake to have a border, say from 12–15 ft. wide, running around the nursery, which may be stocked

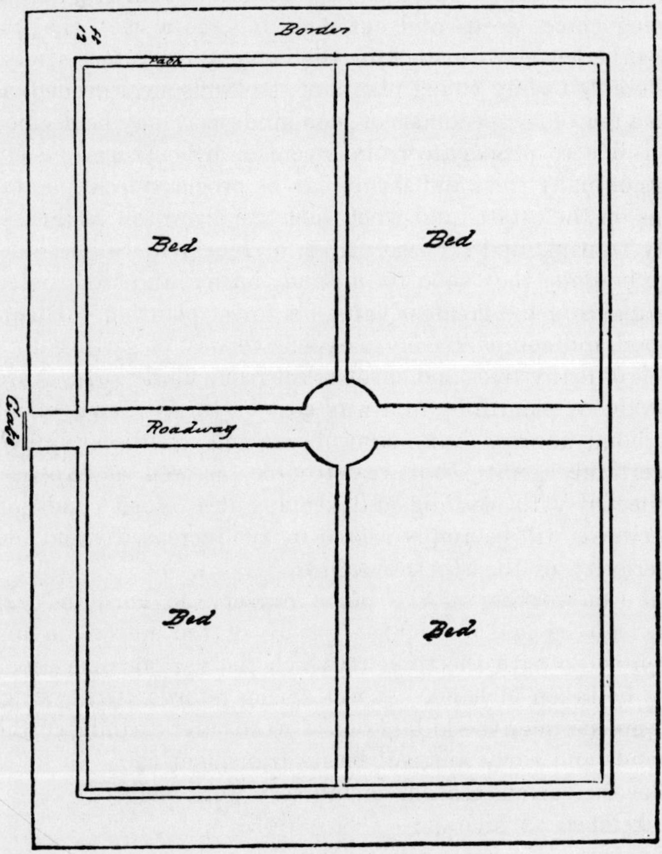

PLAN OF NURSERY.
Roadway, 9 ft. wide; Paths, 4 ft. wide; Border, 12 ft. wide.

with such trees and shrubs as are only limited in demand. A narrow border like this is of great value, too, for planting out seedling stock of the less common kinds, for the insertion of cuttings of the rarer shrubs, as well as for any odds and ends that may be collected.

The site chosen for the seed-beds should be naturally sheltered, or failing this, such artificial shelter as is found necessary should be provided, as exposure of the young plants to cold, cutting winds causes them to become stunted and bark-bound. There should be a few cold frames for raising choice seeds and cuttings, but, as a rule, the less glass the better. In stocking the home nursery, it is always preferable to buy young plants of the kinds most needed, as also a few older specimens of such kinds as it may be deemed advisable to propagate from layers or by cuttings. Seedlings of many trees and shrubs can be procured from plantations on the estate, and when such are grown on, and carefully transplanted for two, three, or four years in the nursery borders, they soon form stout, bushy, and well-rooted specimens of the greatest value for forest planting. This is a good and comparatively inexpensive way of getting up a stock of many trees and shrubs, but particularly such as are reproduced plentifully in a wild state. In the management of a home nursery the amount of care and attention required is certainly great; but any trouble, as well as expense, connected with starting and keeping it in good condition afterwards will be amply repaid by the increased value and superiority of the stock obtained.

In the working of the home nursery no hard-and-fast lines can be laid down, the nature of the season having much to do with the time at which the various operations may be taken in hand. A wet spring retards seed-sowing, a damp summer the killing-out of weeds and cleaning of the ground, and early autumn frosts transplanting.

For the various seasons the nursery-work might, however, be sketched as follows:—

Spring.—By February, all trenching, digging, manuring, top-dressing, and such-like work should be completed. Larch and thorn should be planted at once, as they start early into growth, following up with the various kinds of hardwoods and pines. Layers should be planted out and trees for grafting and budding made ready. Collect larch and pine cones, and when quite dry store away in a cool, airy place, until wanted for sowing. During March and April

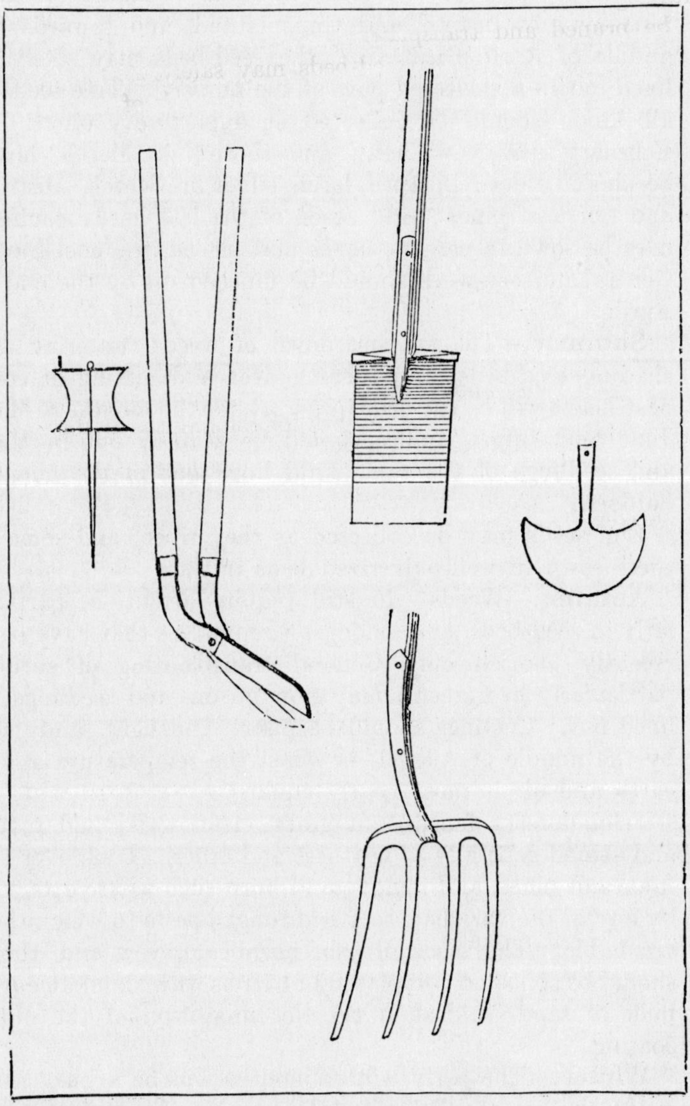

TOOLS REQUIRED IN NURSERY.

general grafting may be taken in hand. Ornamental coniferæ may be pruned and transplanted, and towards the middle of April plants from the seed-beds may safely be lined out in a sheltered part of the nursery. Tree seeds of all kinds should be collected as opportunity offers. In February sow yew, holly and thorn; in March, birch, beech and alder; in April, larch, silver fir, Scotch, Austrian and Corsican pines; and seeds of the less hardy coniferæ may be sown in pans or boxes and placed in a cool frame. General nursery-work should be finished up by the end of April.

Summer.—The keeping down of weeds, watering and shading seed-beds, and turning over and mixing of compost-heaps will be the principal work for the months of May, June and July. Hollies should be planted out in May, and seedlings of the same kind lined out in the nursery borders.

Elm seeds may be collected as they ripen, and some of these sown in well-pulverized beds in June.

Autumn.—Weeds will still require attention, particularly in seed-beds, and amongst young trees that have been recently planted out. General transplanting of shrubs, particularly evergreens, may now go on, and seedlings be lined out. Cuttings should be inserted in light sandy soil by the middle of August, or when the temperature of the earth and air is most nearly equal.

Trim nursery fences, cut grass, clean walks and roads, and attend generally to neatness and order. Look over the brakes of pines, and remove and burn such as are attacked by any of the various insect and fungus pests to which they are liable. The seeds of ash, hornbeam, yew and thorn should be collected and placed in barrels with about an equal bulk of sand, to hasten the decomposition of the outer coating.

Winter.—The early winter months will be a busy time in the nursery, the lifting and dispatching of trees for forest-planting being one of the principal operations—at least, so long as the weather remains mild and open. As time permits, two and three year old plants should be lined out, well-rooted cuttings lifted and transplanted, and layers from old stools carefully cut away and placed in the borders for a year or two before being finally planted out. Turn manure-heaps, and add a small quantity of fresh lime to hasten general decomposition, to sweeten the soil, and deprive the seeds of weeds of their germinative properties.

Sloe, holly, and similar berries may be collected as they ripen, and stored in the usual way.

In November and December, horse chestnut, oak and hazel may be sown in well-pulverized beds of good, rich soil. Ash and hornbeam are sometimes sown in January.

CHAPTER V

FORMATION OF PLANTATIONS

Laying out the Boundary.—Before commencing actual planting operations several preliminaries must be attended to. These will include (1) laying out the boundary, (2) clearing the ground of rough-growing vegetation, (3) drainage where necessary, (4) fencing, and (5) laying out such roads as may be necessary for the efficient working of the plantation. For the benefit of the trees, for shelter purposes, and for the general appearance of the landscape, it would be well were more attention paid to the laying-out of the sites for new plantations. In many cases, however, there is no choice in the matter, the proprietor saying, Here is a field that is of no great value for agricultural purposes, plant it up. But in the case of hillside or moorland planting the matter is usually different, the choice of ground, size of plantation, and method of planting being left entirely in the hands of the forester. Many considerations will tend to determine the position of boundaries—bounds of property, proximity to roads, public paths, and the existence or future probabilities of modes of transit, all being more or less significant factors.

First, however, it is advisable to take into consideration when planting hillside or moorland, the shelter to be afforded to cultivated land in the neighbourhood, but a combination of this with the aforementioned considerations will be all-important.

The form of the outline must also be laid out with due regard to the prevailing wind, and should always present a convex side towards it, as it is obvious that on striking such a curve the force of the wind would be divided and expend

HILLSIDE
WITH PLANTING ARTISTICALLY ARRANGED

SAME HILLSIDE
WITH STIFFLY ARRANGED PLANTATION

D

itself in two opposite directions, thereby losing much of its destructive power. Blending one curve with another should be strictly observed, and continuous straight lines should be avoided, not only for appearance sake, but as it is well known that the most destructive force of the wind concentrates on such outlines.

Woodland Roads.—These are necessary for access to the wood when thinning, the removal of timber and firewood, and for sporting purposes. They should either be pegged out or formed before planting operations have been commenced, and need neither be of an elaborate character nor attended with great expense in construction. According to the area of the plantation, so should the roads be arranged, and a considerable saving is effected by laying out the roads and leaving them unplanted. Usually they are made 30 ft. wide, and only require to have the surface equalised and any stagnant moisture removed by drainage.

It is a good plan, so as at all times to keep the drives in a passable, dry condition, to cut a ditch along each side of the ride, parallel to it, and, say, 18 in. wide by 15 in. deep. The soil so removed will come in for filling up inequalities on the road surface. Steep roads should be avoided by following, in quickly sloping woodlands, the curve of the ground.

For sowing down woodland drives, bridle-paths, etc., the following mixture of grass seeds is to be recommended :—

	lbs.
Timothy grass, hard fescue, tall fescue, and meadow foxtail, 2 lbs. each	8
Smooth stalked meadow and rough cocksfoot, 4 lbs. each	8
Rough stalked meadow and sweet-scented vernal, 4 lbs. each	8
Wood meadow grass	8
Total .	32

In sowing down gravelly or sandy rides, I have found the Lyme or sand grass—*Elymus arenarius* —and the Sea Matweed—*Psamma arenaria*—to answer well, the seeds of each being sown in equal proportions.

Before sowing down woodland rides the soil should be well consolidated, and all surface inequalities levelled.

Fencing.—Local circumstances will often determine the mode of fencing new plantations. Where stones are abundant these may be utilized in the formation of walls, while in districts where slate abounds excellent fences of this kind may be erected at a cheap cost.

Turf walls and ditches are sometimes formed as plantation boundaries, but unless carefully looked after they are rarely very successful.

Stone and turf walls are now, however, almost entirely superseded by iron fencing, it being not only readily conveyed to any desired point, but quickly erected and moderately cheap.

To recommend any particular system of wire fencing would, for various reasons, be out of place, for amongst the numerous kinds now offered almost any desired pattern can be obtained.

In any case the fence need not exceed 3 ft. 6 in. in height, with seven-strand wires, two No. 6 and five No. 7 ; and, as this is placed in position at so much per mile by the makers, it is always well, so as to avoid bad erection and keep down expense, to have it so put up. Where an abundance of larch timber is growing on the estate, it would, perhaps, be unwise to employ iron, but in the majority of cases it will be found cheapest in the end to have the fences erected wholly of iron and wire.

The straining-posts should be of sound, well-seasoned larch or oak, and 7 ft. long by 6 in. square, or if round 7 in. in diameter.

The intermediate posts are usually of larch, 5½ ft. long by 3½ in., by 3 in., or if round, not less than 3½ in. diameter at small end.

The straining-posts are placed in the ground at 150 yards apart, with an additional strong post at every angle or curve that occurs in the line of fence. The intermediate posts are pointed and driven into the ground along the line of fence at 6 ft. apart.

In order to make a fence proof against the inroads of

sheep and cattle not less than six wires should be erected, and galvanized strand wire is preferable. The total height of the fence need not exceed 3 ft. 6 in., and the top wires should be placed farther apart than those lower down.

In tightening the wires a straining machine is usually employed, but brackets for the same purpose are preferable, these being attached to the straining-posts.

They are of particular value, as the wire can be tightened or slackened at will.

The posts should be rounded on the top or sawn with a slant, so that the rain may run off.

Draining the Ground.—Efficient drainage must be considered as one of the most important operations in the formation of a plantation. Every portion of the ground may not, probably, require to be drained, but where it is at all surcharged with moisture the removal of such will be a step in the right direction if the future welfare of the trees is a point of first importance. In commencing draining the position of the main outlet must first be determined, and in doing so every fall of the ground should be taken advantage of. It is quite impossible to lay down rules as to the number, sizes, and distances apart of the various drains, these being points that can only be satisfactorily settled on the spot, and when the nature of the soil, lie of the ground, and amount of rainfall to be carried off are determined. In most cases, however, the main drains should be from 2 ft. 6 in. to fully 3 ft. wide at the top or surface, from 10 in. to 12 in. wide at the bottom, and about 3 ft. deep. The minor drains may be less in proportion to the mains, and are usually 2 ft. deep, 2½ ft. wide at top, and the width of the draining spade at bottom.

According to the nature of the ground so will the distances at which the drains should be cut vary, but in most cases from 15 ft. to 30 ft., sometimes even less, and sometimes considerably more. The minor drains should never run at right angles to the main, but at about 45°, which will prevent the mouths becoming choked up when there is a rush of water in the main. The soil removed from the drains should be evenly spread out over the ground-surface.

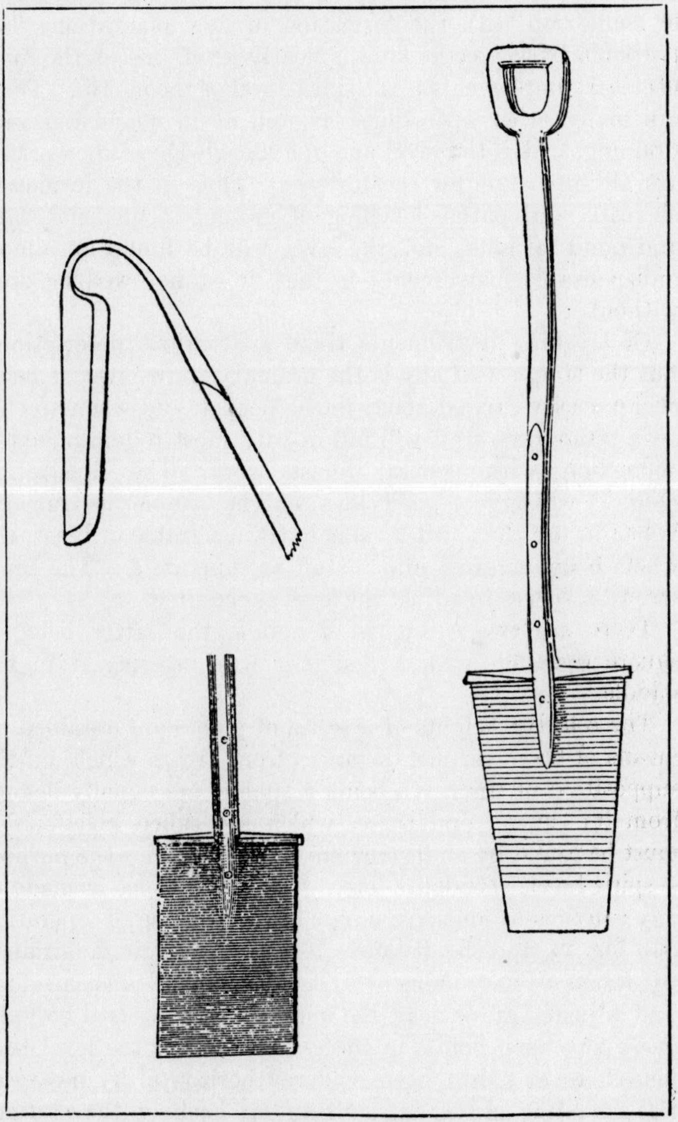

TOOLS USED IN DRAINING

Levelling.—In the carrying out of drainage operations in connexion with the formation of new plantations, but particularly where the land is nearly level, one of the most necessary implements is the spirit-level or theodolite. There are many other operations, as well as in connexion with draining, where the level and the knowledge of its working are all-important for the forester. Thus in the formation of roads and paths, levelling of ground inequalities, lake and pond forming, etc., the level will be found an almost indispensable instrument—in fact, it cannot well be done without.

Of levelling instruments there are various descriptions, but the simplest of any is the ordinary spirit-level, it being cheap, easily carried about in the pocket, and when erected on a temporary staff will fall in with most requirements in connexion with general forest work. The illustration (Fig. 1) will give a good idea of the simple instrument. When in use the spirit-level is fixed in a frame of brass, the whole being screwed into a staff or support, *e*. The brass screw, *d*, serves to adjust the level as required.

There are two eyesights, *a* and *b*, the latter being a square opening, with a fine hair wire crossing it in the middle.

The relative heights of a series of points are obtained by means of their vertical distances from others which, on the supposition of the earth being a sphere, are equally distant from its centre, and these, which are called level-points, must be found by an instrument constructed for the purpose —spirit-level, theodolite, etc. Generally choice is made of any convenient stations, *a*, *b*, *c*, *d*, on the line of operation (see Fig. 2), and the distances between them are determined by actual ad-measurement. The instrument is then set up and adjusted at, or near, the middle of the interval between every two such points in succession. When the level thus placed, as at *f*, has been rendered horizontal by means of the adjusting screw, an assistant at each of the stations *a* and *b*, holding what is called a station-staff in a vertical position, moves an index along the staff, up or down, as dictated by the observer of the level, till it coincides with the intersecting wire as seen in the eyesight or telescope. The points thus determined on the stave are represented by *e* and *g*, and these are termed level-points, or points equally distant from the centre of the earth. Therefore the heights

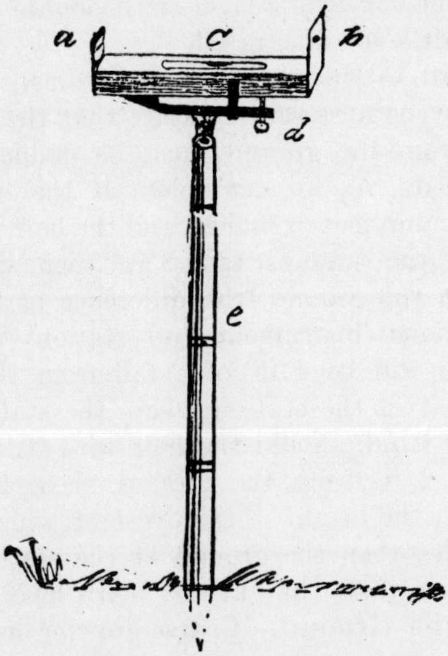

FIG. 1. SPIRIT LEVEL.

a, e, and *b, g,* being read on the graduated staves, the difference between them will give the relative heights of the ground at *a* and *b*. Similar processes are repeated with respect to the points *b* and *c*, and *c* and *d*, the instru-

FIG. 2. METHOD OF LEVELLING.

ment being placed at *i* and *m*, midway between them. Usually the heights *b g*, *c l* and *d n* are inserted in a column headed "Foresights," and the heights *a e*, *b h*, *c k* and *d n* in a collateral column headed "Backsights."

The difference between the sums of the numbers in these two columns will be equal to the height of one extremity (*a*) of the line, above the other (*d*). When a number of levels have to be taken in succession, it will be found a saving of time to use the surveyor's level or theodolite instead of the ordinary spirit-level illustrated.

With regard to the use of the instrument as shown in Fig. 1, it may be necessary to state that the height of the eyesight *b* from the ground must be deducted from the point observed. As an example: if the object-pole or staff be marked in feet or inches, and the hair wire in taking a sight strike the same at, say, 8 ft., then, if the eyesight be 4 ft. from the ground, the difference of level between the two stations (instrument and station) will be 4 ft., that is, there will be 4 ft. of a fall from the spirit-level station to that of the station where the staff was placed. On the other hand, should the hair wire strike the object pole at, say, 2 ft. from the ground, these 2 ft. must be deducted from the height of the eyesight, which, as we have said, was 4 ft., then the ground at the station-pole must be 2 ft. higher than that at the spirit-level.

Clearing the Ground.—Coarse-growing herbage, which often includes heath, gorse, and rough grasses, that would interfere with planting operations, should be removed, but not indiscriminately, as it may prove of inestimable value under certain conditions. On exposed and high-lying ground, the heath and grasses will prove of great advantage to the young trees, and usually they are not too luxuriant or apt to cause damage at high altitudes. Where, however, the growth of such shrubs or grasses would be detrimental to the young trees, by all means have them removed. The best way is either to cut them over or grub out by the root, and burn on the ground. During a continuance of dry weather it may be possible to burn without either cutting or uprooting, but, generally speaking, removal of the roots is to be encouraged.

Pitting.—The advantages of pitting over any method of planting cannot be questioned, and this is particularly the case with ground that has hitherto been uncultivated. Compared with notch-planting, this system is, no doubt, more expensive, but that the future benefits to the plants are greatly enhanced is admitted by all practical arboriculturists.

It is not, however, to be inferred that, though pitting is preferable, notching is to be abandoned, for there are many precipitous, rocky places where it would be the only practice feasible, and plants so inserted have often succeeded admirably.

The pits should in all cases be opened for some time before planting—indeed, in unfavourable soils and situations, it is a good practice to have such work performed in autumn and the plants inserted the following spring. By so doing, the earth that has been removed from the pits will lie fully exposed to the mellowing influences of frost and sunshine, so that when the time for replacing it comes round, it will be in the best possible condition for applying to the roots of the young trees.

The pits in uncultivated lands should be made circular, about 18 in. in diameter and fully 12 in. deep, and the sides and bottom well loosened up with a pick. In loose or recently cultivated soil the pits may be much smaller. Take off the surface turf in halves, placing these on one side of the pit, and the soil on the other, for ease and convenience in planting. When the ground slopes quickly the soil removed from the pits should, so as to facilitate quick replacing, be deposited on the higher side.

Planting.—This may be successfully carried out during all open weather from about the end of September to the beginning of April, but, generally speaking, autumn planting is to be recommended. There are several exceptions however: such as when we have to deal with peat bog, water-logged soil, exposed hillsides, or land by the sea-coast. Where the newly-inserted plants have to cope with prolonged storms, such as we get on hillsides or by the sea, or contend with very uncongenial soils, it is always a wise policy to defer planting until spring, or just when the trees are about to make a start to growth, as they, with their freshness and vigour undiminished by the change from the nursery border to the more trying surroundings just referred to, are more likely to take hold at once and succeed.

The battering and swaying that autumn planted trees receive when exposed to the hurricanes of our hillsides or seaside sites so enfeeble them that, in spring, when growth should commence, the majority will be found to be in a very unsatisfactory state, whereas, by inserting in spring, when growth will soon be at its full activity, the chances of succeeding are greatly enhanced. In peat bog the antiseptic properties of the soil act dangerously on the roots of young trees if allowed to remain therein for some time before active growth has commenced. However, with the exceptions cited, tree planting throughout the British Isles generally should be taken in hand as soon as the leaves of the hard-wood species have fallen, which usually takes place about the second or third week of October, much depending on the particular season. Lift the plants very carefully from the nursery brakes, and do not, on any account, tolerate the too-often-enacted practice of tearing the trees from the ground, and before they have been properly loosened on both sides of the lines with a fork. To lift nursery stock properly—and the extra expense incurred in so doing is money well spent—a trench should be thrown out along each side of the line and the soil undermined from beneath the roots, so that the plants can be lifted without tearing or straining the tender rootlets. It is not important, if the plants are inserted soon after being lifted, that soil should accompany each, the roots being plentiful and unmutilated making up for the want of this. Where, however, the plants have not to be conveyed far from the home nursery to the plantation there is no need to remove much of the soil, for if left intact the young tree is far more likely to start away freely into growth than if this was shaken clean off.

Immediate planting after being lifted is to be strongly recommended, the evil of allowing plants to lie about exposed to wind and weather being well known. Should it, however, not be convenient to plant at once, the nursery-stock should be stood closely together, and some damp straw,

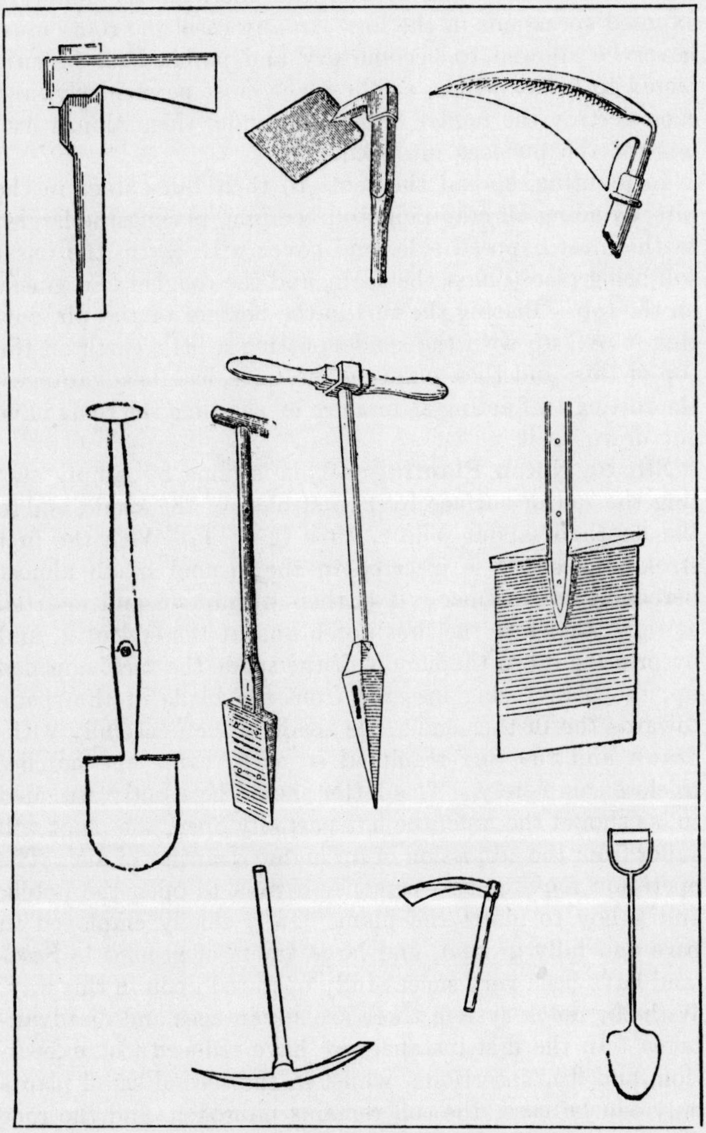

TOOLS USED IN PLANTING

leaves, or soil heaped around the roots of the outer or exposed specimens in the lot. In any case, the roots must never be allowed to become dry and parched, or be subjected to frosty winds, as these act most perniciously, and soon destroy the tender fibres and render them almost useless for the purpose intended.

In planting, spread the roots to their full extent in the pits, avoiding all cramping and bending, placing the largest to the most exposed side, and cover with earth, the finest soil being placed next the roots, and the rougher and grassy on the top. Placing the turf in the bottom of the pit, cutting it well up with the spade, putting a little earth on the top of this, and then planting the trees, has its advantages, the rotting turf acting as manure by the time the roots have got down to it.

Slit or Notch Planting.—This is done by simply cutting the sod or surface by two strokes of the spade, and to the depth of about 5 in.: thus L or T. With the first stroke the spade is inserted in the ground in an almost perpendicular manner; it is then withdrawn and inserted at right angles to the first notch and at the end of it, and by pressing down the handle of the spade the turf is opened up, the plant being inserted from the blade of the spade towards the further end; the spade is then carefully withdrawn and the turf trampled so as to cause the notches to close completely. This latter should be strictly attended to, as should the notch be left partially open, the plant will suffer from the admission of an undue quantity of air. The operation requires two persons—a man to open the notch, and a boy to insert the plant. It is chiefly employed in bare and hilly ground, and large tracts of ground in Scotland have been very successfully operated upon in this way. With the notch system there are advantages and disadvantages. In the first instance, we have reduced cost, expedition, and firm insertion; whilst in the second small plants only can be used, the soil remains unbroken, and the root system presents an unnatural position.

The plants used with the notch system should not exceed 9 in. in height.

The Planting-iron has been found of great value for inserting small forest trees in rocky ground, where it would be almost impossible to do so in any other way. It is 17 in. long, weighs 3 lbs., and can be conveniently used with one hand.

Holding the "iron" slackly, the planter strikes it into the ground with a force sufficient to drive the sharp, heart-pointed blade in about 3 in. or 4 in. By pressing it down and towards the planter, with a slight twist to the right, the left corner of the turf is opened up, the plant being carefully inserted with all the roots beneath the ground. The iron is then withdrawn, and the loosened turf made firm by tramping.

Only a small plant should be inserted with the planting-iron, and great care taken to insure the soil and turf being firmly pressed around the stem of the young tree.

Planting Trees too deep.—Fully 50 per cent. of the deaths amongst newly-planted trees may be directly attributed to the pernicious practice of too deep planting. Cases come before us frequently where, owing to burying the roots of the trees and shrubs at too great a depth in the soil, they have either died out or gradually become unhealthy owing to strangulation and want of air. It is a mistaken idea that by placing the roots at an excessive depth in the soil the young tree will be better enabled to withstand wind, and so remain in an erect position. Most workmen quite ignore the original mark on the tree stem as to the depth it stood whilst in the nursery, and go to an opposite extreme by covering up both root and stem to an unnecessary depth. Large trees suffer quite as much from being planted too deep as those of smaller growth. Quite recently we were asked to look at a large number of lime trees about 16 ft. in height that had been planted for several years without a sign of upward growth. The reason for this was not hard to discover, for on unearthing one of the trees it was found that the roots were covered with fully 2 ft. in depth of soil. The trees in question had never budged an inch, and were from year to year gradually on the decline owing to the ruinous practice of too deep planting. No better guide as to the depth at which a tree should be planted can be had than the distinct mark on the stem showing the depth at which it stood whilst under nursery management. In order to keep newly-planted trees from being damaged by rocking with the wind, and to preserve them in an upright position, firm trampling is all that is required. In very exposed situations it may sometimes be found necessary to place a small piece of turf against the stem on the opposite side from which it is leaning, and to make this firm by means of the foot. Staking, at least in the case of small transplants, is never necessary, though in the case of larger trees and where the situation is exposed, light poles thrust into the ground to which the stems are made fast may be found necessary. Of course, it is a mistake to allow trees to rock about in the wind to such an extent that the roots become strained and barked, and holes worked around the stems, but this can readily be prevented by periodical examinations of the newly-planted trees. Stones should never be placed in the holes referred to, as they damage the roots, a piece of stiff turf or heavy soil being far preferable. With a large experience of tree planting I find that the general tendency is to plant too deep—an evil that it is hoped these notes will be the means of remedying, in some cases at least.

CHAPTER VI

TREES FOR ECONOMIC PLANTING

From a commercial point of view the best trees to plant are such as will produce the largest volume of the most valuable timber in the shortest space of time.

Amongst the several hundred species of trees that are cultivated in this country it is a remarkable fact that less than twenty can be recommended for profitable planting. Of course many others are valuable as shade and shelter producers, but from a strictly economic point of view, that is to say where the value of the timber is a point of first consideration, only about ten hardwooded and eight coniferous species can be included in the list.

Amongst hardwoods we have the oak, ash, beech, sycamore, elm, Spanish chestnut, willow, poplar, alder and birch; while coniferous trees would include the larch, Scotch and Corsican pines, common and Sitka spruce, silver and Douglas firs, and possibly the giant Arborvitæ and Weymouth pine.

The **Oak.**—Than the timber of the oak none other produced in this country is more lasting or valuable—indeed, in few other trees are the qualities of strength and durability, hardness and elasticity combined to such an extent. As might be expected, however, the value of oak timber is greatly influenced by soil, that produced on deep, heavy loam or clay lands being of superior quality. The colour of the wood also varies greatly with age from a dark grey to a deep rich brown and even green. Brown oak, which is almost confined to the Midland counties of England, is considered to be wood in the first stage of decay, and attacked by a fungus, and is generally associated with trees that are long past their prime and whose larger limbs have become hollowed and unhealthy. The finest examples I have seen of brown oak were produced at Ampthill in Bedfordshire, and on the Welbeck Estate in Notts. The beautiful vivid green colour assumed by oak timber under certain conditions is due to the action of the fungus *Peziza æruginosa.* As this colouring matter in the timber is quite permanent and cannot be destroyed, wood so affected is eagerly bought up by the makers of fancy furniture. Unhealthy or dead trees in damp shady positions are most often attacked by the fungus, and we have known timber of the oak when left lying about in the woodlands to assume this colour. Owing to the scarcity and value of green oak timber experiments have been undertaken to produce the colouring by artificial means, but as far as we are aware the results were not satisfactory. The best examples of green oak timber that we have seen were produced in Kent, and in the North of Ireland.

The **Ash.**—When viewed in a purely economic sense the ash must, next to the oak, be considered as our most valuable forest tree. The massive, deeply fluted, or cylindrical trunk, the weighty swelling branches, and the usually pendant masses of the freshest pea-green foliage, all combine to render this tree one of the most majestic for ornamental planting, but particularly so when associated with others that are of a darker shade of green. The ash is indigenous to Europe, Northern Africa, and North America, while throughout the British Isles it is widely dispersed.

It thrives tolerably well in most soils and situations, but the finest timber is produced in fairly sheltered sites, and where the soil is rich and open or freely interspersed with loose rock or stones. In order to produce clean, springy timber, such as is in request for the making of tool handles, aeroplanes and agricultural implements, the ash should be planted thickly in order to induce straight stems that are for the greater part destitute of branches. For black, peaty soils, or that of a dampish, loamy character the sapling ash is peculiarly suited, in which, after being cut over, it will reproduce itself freely. At from thirty to forty years' growth the ash is most valuable for handle wood and for agricultural implements.

The timber is hard, heavy, and flexible, and though rapidly grown it is tough and elastic above that of any other tree grown in this country, hence its universal employment for machinery and other special purposes where great strength, combined with yielding powers, are points of first consideration. Few timbers become useful at such an early age as does the ash. At from four to six years' growth it is fit for walking sticks, for spade handles at ten years, while after that age the uses to which it is applied are very numerous and diverse. For the cartwright, the agricultural implement maker, carpenter, cooper and turner, it is of special value. No timber has been found to equal it for the making of oars, pulleys, blocks, ladders, hoops, and crates. Owing to its sawing without splintering it is used for milkpails, kitchen tables, staircases, and similar purposes, while when of large size it is greatly in demand for furniture making.

For smoking herrings ash wood has a certain local demand. Potash is procured from the ashes of the branches, and the bark has a special value in the tanning of nets.

The **Beech.**—For profitable planting the beech is one of our most valuable trees, while it has the extra recommendations that it grows rapidly, succeeds well in the shade, and reproduces itself freely.

The finest beech timber is produced on chalky or deep sandy soils, the former being preferable as may be instanced by the comparatively high price obtained for that grown on the Chiltern Hills and chalky formations of parts of Kent and Herts. The growth of the beech is rapid and it attains its prime at from sixty to seventy years, after which age the timber often becomes black at the heart and accompanied by rottenness and shakes which render it almost useless for constructive purposes. It is better suited for indoor than outdoor use and is extensively used in the making of Windsor chairs, for millwright and engineering purposes, for the handles of carpenters' tools, for bobbins and, when clean and of large size, for the calendar machines of cloth manufactures.

Growing on comparatively poor soils where other species

STANDARD BEECH TREE

will not produce profitable timber the beech is particularly valuable on account of its reproducing itself freely from seed and withstanding a greater amount of shade than perhaps any other species. It associates well with the oak under cultivation.

The **Sycamore** for profitable planting can be recommended owing to the high price which can be obtained for timber of large size, though even when of thirty years' growth it finds a ready market for bobbin making and other turnery purposes.

The most suitable soil for the sycamore is a dry sandy loam or even deep sand. The tree is peculiarly suitable for planting in exposed or seaside situations and also for standing alone as a field or hedgerow tree. The timber is white, smooth and free from grain. It is used for curtain rings, churns, butter prints, backs of violins, founders' patterns, cutting boards, and in the making of wooden vessels and furniture. For calendar machines it is especially valuable.

The **Sweet** or **Spanish Chestnut** is largely cultivated for the value of the timber it produces. It thrives best on deep gravelly soil or on granite or sandstone with a dry subsoil, and with a southern or western exposure. The timber is of excellent quality and lasts well either in water or above or below ground. It is often substituted for that of the oak, and is extensively employed for mill timber, bridges, fencing of all kinds, posts, stakes, casks, and for hop poles.

The Spanish Chestnut is a tree of rapid growth and when the timber is of fair size it finds a ready market, though at a lower figure than that of the oak.

Unfortunately the timber produced on certain soils and situations is much reduced in value owing to cup shake.

Elm, both English and Scotch, are extensively planted for the value of timber produced.

The timber of the English elm is of a rich brown colour, hard, tough and usually twisted in the grain. Owing to the strength of its lateral fibre the timber is much in request for making blocks for ship's riggings, while it furnishes naves for wheels, coffin boards, furniture, pumps, piles, and is employed largely by the wheelwright. For using under water it is the best wood cultivated in this country.

The wood of the Scotch or Wych elm is considered to be

more valuable than that of the English species, and is used for similar purposes and also as a substitute for ash. It is very liable to become rotten at the core.

The Alder and its Uses.—Whether viewed from a strictly commercial standpoint, for its justly recognized capacity for thriving luxuriantly where few other trees could exist, or for its ornamental qualities, the common alder will be found to rank high among British forest trees. It usually attains to a height of from 50 ft. to 60 ft., with a stout, well branched trunk that is more or less gnarled and fluted. The branches have an upward inclination, and are well clothed with roundly wedge-shaped leaves that are of a deep, dark green colour. With age the bark usually assumes a dark brown, or almost black, colour. The alder is found in all parts of Europe, North Africa, and from Asia to Japan. For planting in wet, even swampy, situations, where only a very limited number of trees could eke out an existence, the alder is of particular value, and it is in such situations that it attains to the largest size and produces the greatest quantity of timber. Even in situations where the poplar and willow find a difficulty in battling with the excessive and stagnant moisture the alder thrives luxuriantly and reproduces itself freely from seed. As an ornamental tree the alder is not much in request, although the stately habit and ample deep green foliage render it of value in that respect. The fine old specimens by the river and stream sides at Esher Place in Surrey, have a beauty that is quite their own, and it is in such tranquil situations that the tree is seen to perfection—the beauty of form and picturesque appearance causing it to be unrivalled in river and lake scenery. The tree is readily propagated and comparatively cheap in consequence, grows rapidly after becoming established, and is not subject to disease or premature decay.

Though soft, the timber of the alder is in much request, and in consequence numerous wants are supplied by it. It is very durable under water, and for this reason is largely employed for piles, bridge foundations, water pipes, and for lining the banks of rapid running streams and rivers. The celebrated bridge of the Rialto, at Venice, was, according to Evelyn, built on piles of alder wood, while the city of Ravenna was stated to have likewise been built on piles of the same wood. For herring barrel staves the wood is also in request, as it is for mill bobbins and turnery work. In cabinet work and for cheap furniture alder wood is largely used, while as it does not readily split, boards for the bottoms of carts and wheelbarrows are frequently made of the same wood.

The beautiful pale pink colour which the timber permanently retains causes knotty planks to be in great demand for veneering purposes.

In Wales, and throughout the Midland counties, large quantities of alder timber are consumed in the making of clog soles, which, after being roughly formed in the woodlands where the trees have been felled, are sent by rail to several of the Lancashire towns to be finished off. Throughout Ireland—chiefly the north—large numbers of clog soles, made of alder, are annually employed in the manufacture of cheap boots ; indeed, in many parts the making of clogs is quite an industry.

The bark of the alder is used in tanning leather, though in much smaller quantities at present than was the case half a century ago, when oak bark fetched as much as £10 a ton, and when none of the chemicals that are now so commonly employed were offered in the market. Excellent gunpowder is made of the wood, said to be second only in quality to that prepared from the dogwood. The young shoots, according to the peculiar way in which they are prepared, are employed in dyeing red, brown, yellow, and black.

Alder is generally in good demand at all stages of its growth, and is seldom grown to very large dimensions. For the clogger, turner, or charcoal burner it is of greatest value up to about thirty years' growth, while by cutting it over at that age a second crop springs rapidly from the stools.

The **Birch.**—It is hardly necessary to speak of the hardiness of the birch, for no other native tree, not even the Scotch pine, ascends to such elevations in Britain. The higher the tree ascends the more shrub-like it becomes, until on very exposed rocky sites at great altitudes it hardly exceeds a yard in height.

As regards soil, the birch is by no means particular, for we find it succeeding well even on that of a poor rocky or gravelly character. The largest individual specimens always, however, occur at not too high an altitude, and on soil of a light, loamy nature, an abundance of moisture being still further favourable to quick development. For planting on exposed ground for shelter-giving purposes, but particularly where the soil is thin and poor, the birch is a most valuable tree. Its growth is fairly rapid, and it does not suffer to any appreciable extent either from insect or fungoid attacks. As a coppice tree it is also valuable.

The timber of the birch, though of little value for general estate purposes, is largely employed in the arts and manufactures. It is much used for thread bobbins, turnery work, shoe pegs, hoops, and fish barrels, while it makes excellent firewood, and yields superior charcoal for smelting purposes. In Wales large quantities of birch wood are cut into clog soles, while the sabots used by the French peasantry are also made of that wood. When of large size and good figure, furniture and veneers are made of the wood, and on the Continent felloes for carriage wheels. Brooms and switches are made of the smaller branches or sprays, while the bark is used for tanning, and an oil expressed from it is employed in the preparation of the well-known and fragrant Russian leather.

The **White** or **Huntingdon Willow** (*Salix alba*).—Whether for ornament or utility the White or Huntingdon Willow must be considered as one of our most valuable trees. The timber sells readily at all prices up to 10*s.* per cubic foot, and when it is considered that the tree will succeed in dampish ground where most other species decline to grow, and that it is of rapid growth, attaining maturity in less than forty years, its value in economic planting will be recognized. At the outset it may be well to point out that the wood of the particular willow from which the best class of cricket bats are manufactured sells at a higher rate than any other timber that is cultivated in this country. There are many kinds of willows found growing throughout the British Isles, but one alone produces the particular class of wood from which first-class bats are turned out. Until quite lately the timber of the White or Huntingdon Willow

STEM OF THE TRUE "CRICKET-BAT" WILLOW (*Salix cærulea*)

(*Salix alba*) was largely used in the making of cricket bats, but it has been found that a cross between that species and the crack willow (*S. fragilis*), and named *S. cærulea*, produces by far the best wood for the manufacture of high-grade cricket bats. In the trade the "Cricket Bat Willow," as it is now known throughout England, is popularly designated as the close-barked willow in order to readily distinguish it from the open-barked or crack willow. Confusion generally exists in determining the various forms of willow, but in *S. cærulea* the branches incline upwards; indeed, the tree has a semi-fastigiate form of growth, and the branches also have an upward tendency. The bark is a dark grey, with long, straight, narrow fissures closely arranged and from which the term "close-barked" is derived. The leaves are of a bluish tint or covered with bluish-grey hairs beneath and long and narrow in shape, while an unerring point of difference between the hybrid and other willows is that the tree produces only female flowers. It may be well to mention in connexion with the crack willow that the bark fissures are far more rugged and placed farther apart than is the case with the true cricket bat willow, *S. cærulea*. The great importance of recognizing and growing for purely economic purposes the true variety will be apparent when it is mentioned that makers of cricket bats will have nothing to do with any but the true "close-barked" tree, and the English bat-maker is keen to recognize the characteristics of the timber he requires, and will not stick at paying exorbitant prices for trees of the right kind.

As showing the value in England of the timber of the true bat-making willow, it might be mentioned that in many instances that have come under my notice as much as 16*s.* per cubit ft, or six times the price of the best oak, has been paid for trees of the true *S. cærulea*. A single tree growing in London lately sold at £10, and in Hertfordshire eleven trees fetched the handsome sum of £81, while £20 was refused for four trees growing in a wood in Essex. Such prices are, however, exceptional, though on a visit to two of the largest bat-making establishments in the metropolis I was told that for several years past the average price paid for willow worked out at 6*s.* per cubit foot. A well-known grower tells us that if planted in suitable soil a "set" or cutting, which usually costs 1*s.* 6*d.*, will in fifteen years be worth about £6.

The home of the true bat-making willow is in the Eastern English counties, and it is there that the manufacturer goes when wishing to purchase the most valuable timber for his special work. The propagation of the bat-making willow is simple, either by rooted cuttings or "sets," the latter being the best and cheapest method. "Sets" are usually from 12 to 20 ft. long, with a basal diameter of, say, 3 ins., the best being got from pollarded trees, and straight, clean, branchless shoots are preferable. They should be stripped of all side branches for about three-fourths of their length and only cut in spring. By placing these sets together in a ditch or pond for about a month rootlets will be emitted, when they may be planted in previously-prepared holes, which are formed by driving an iron rod or stake in the ground for a distance of, say, 2½ ft.

The **Poplar.**—Several species of poplar are valuable for the timber they produce, as also for their rapidity of growth and succeeding in low-lying, damp ground. When clean grown and of large size the timber sells readily at prices which vary from 1*s.* to 1*s.* 6*d.* per ft.

Probably the most valuable species is the white poplar (*Populus alba*), though the black Italian (*P. monilifera*) produces excellent timber for which there is generally a demand.

From a purely commercial point of view the above hard-wooded trees are the only kinds that can be recommended for planting in this country. The timber of the lime, hornbeam, walnut and one or two others at times sells at a fair price, but the small quantity offered shows that they have not been considered worthy of attention where the economic value of the plantations was being considered.

Coniferous Trees

Amongst the many conifers that have been introduced to this country during the last century very few can be recommended as suitable for profitable planting. The following, so far as is known, are the only species to be recommended :—

The **Larch** (*Larix europœa*).—Both for and against the larch much has been written and said, particularly of late years ; but, however much has been said in commendation of it, there can be no doubt that to overstate its value as a timber tree in the economy of British forestry would be a task of some difficulty. When we combine its great, almost peculiar, aptitude to suit itself to nearly all conditions of soils, altitudes, and diversities of climate, its long-established value as a timber tree, rapidity of growth and ease of culture, it is clear that no other tree cultivated in this country can be ranked on a par with the larch. Unfortunately, however, of late years the larch has suffered much from disease, the predisposing causes of which may mainly be attributed to the variableness of our spring weather, and the rapidly degenerating state of the tree—the latter chiefly brought about by an injudicious selection of seed. By far too little attention has been paid to this important matter, the result being that weakness and tenderness have got into the constitution of the tree, and it is thus unable to withstand even a few degrees of frost. So weakened, blight, fungus, and ulceration find a footing, and thus the fell disease is generated about which so much has been said and written of late years. In my own opinion, strengthened by careful investigation and research, induced tenderness in the constitution of the larch is the primary cause of disease, cold winds and frost being the destructive agents, and ulceration the direct consequence.

Injury to the roots of the larch in transplanting is attended with most injurious results. In corroboration of this, it may be stated that natural or self-sown trees are, in this country, almost exempt from disease.

The variableness of our spring weather is, no doubt, one of the predisposing causes of disease, for although no degree of cold experienced in this country can injure the tree when leafless, yet few are more sensitive when in young foliage.

The durability of the wood of the larch is well known, and this peculiarity is even noticeable when of only a few years' growth. As compared with Scotch and spruce firs, the wood of the larch is about twice as durable—a fence of the latter cut at from twenty to thirty years' growth lasted from seventeen to twenty years, while that of the spruce lasts about eleven years, and the Scotch seven years. This refers especially to rails, not to posts, which decay in about half that time. For mining and railway purposes the durability of larch wood makes it much sought after, this being further enhanced by its extreme lightness, the weight of a cubic foot when seasoned being only 34 lbs. It takes a beautiful polish, works with great freedom, and, when fully seasoned, is not at all liable to twist or warp.

Substitutes for the larch have often been recommended, but, in the true sense of the word, there are practically none, although, doubtless, some of those whose claims have been set forth might to a certain extent reflect one or more of its valuable qualities.

Sitka or Silver Spruce (*Picea Sitchensis*).—From a purely commercial point of view the Sitka or Silver Spruce is probably the most valuable of any of the family to which it belongs. It is a tree of noble growth in this country, several specimens being well over 100 ft. in height and with boles that girth fully 10 ft. at a yard from the ground, these dimensions being attained in seventy-five years. When used as a standard, the tree is one of great beauty, the stiff and rather irregularly disposed branches being thickly beset with vivid bluish-green foliage that is silvery on the under side. It delights in a cool, moist loam and not too exposed situation, but when grown on gravel or any warm soil the foliage is distinctly meagre and affected by red spider. The timber, which is remarkably light for its bulk, strong and flexible, is of great value in the making of aeroplanes, and special logs find a ready market at highly remunerative prices in the London market.

The **Corsican Pine** (*Pinus laricio*).—This is, undoubtedly, one of the best all-round conifers that has found its way into the British Isles. It is of very rapid growth, and well suited for planting, even in the most exposed and windswept situations; a non-fastidious subject as to soil, and a valuable timber producer.

As to its adaptability for withstanding long-continued cold blasts at high altitudes, ample evidence can be adduced on many an English and Scotch estate where the pine has been introduced into the woodlands in such quantity as in certain places to form the ultimate or standing crop. In North Wales, on one of the Snowdon range of hills, I have planted the Corsican pine in great quantity. The plantation was, for the greater part, fully exposed to the dreaded south-westerly wind, which at times blows hard and long, and sweeps the hillsides with terrific fury; yet, under such conditions, the Corsican pine has done remarkably well—in fact, proved itself to be well suited for planting at high altitudes on our English hillsides. Even at the highest point of the woodlands in question, this pine has thriven in a manner that is quite surprising, and thrown its stoutest branches out into the very teeth of the blast, and that where hardly a hardwood tree could survive, and even the Scotch fir shrank from the cold and almost unceasing storms. Other notable instances of how well the Corsican does on exposed ground and high altitudes might be pointed out—such as at Blair Athol, in Perthshire, at 700 ft., where it is thriving amazingly; and again in Yorkshire, one of the most barren and wind-swept of English counties, where in parts, it grows with a luxuriance that is almost unparalleled in any other part of Britain. The timber produced by the Corsican pine in this country is strong, tough, elastic, very resinous, and easily worked; this is speaking of trees of fully fifty years' growth. It thrives well on gravelly soil, some of the largest specimens of the tree in this country growing along the margin of a disused gravel-pit.

It may be said that the Corsican pine is perfectly hardy, peculiarly well suited for planting in exposed situations, a rapid and valuable timber-producer, a tree that is cheaply and easily raised from seed, and one of the most non-exacting conifers as regards choice of soil that could be named—all qualities of the highest value in a timber-producing tree and a combination that is rarely found in any other species.

In France extensive plantations of the Corsican have been formed, while it has also been introduced extensively into the State forests by the Prussian Government.

The Weymouth Pine (*P. Strobus*), whether viewed in an ornamental or economic aspect, must be considered as

THE WEYMOUTH PINE AT GWYDYR CASTLE

another of the most valuable pines that have been introduced into this country. Admit, we must, that in certain situations the cultivation of this handsome tree has been attended with no very promising results ; but then it should be remembered that, like most other trees, the Weymouth pine has its likes and dislikes of soil, as well, indeed, as of aspect and altitude. That it has succeeded well, and produced an unusually large quantity of clean and firm wood in various parts of the country cannot be denied ; but then in such places its peculiar wants have been attended to. At Gwydyr Castle, in North Wales, the tree succeeds admirably, specimens fully 90 ft. in height, straight as arrows, branchless for three-fourths their length, and fully 8 ft. in girth at breast-high, being not uncommon. The soil is rocky debris, largely intermixed with vegetable refuse, fairly moist at all times, but without stagnant moisture.

On the western borders of Ross-shire, at Strathkyle, where the altitudes vary from 100 ft. to 1,200 ft., the Weymouth pine is making rapid progress. We do not wish it to be inferred that the Weymouth pine alone is suitable for planting at high altitudes and on exposed situations ; but that it will grow rapidly and produce useful timber in partially-sheltered districts has been proved by those who have paid particular attention to the value of the more recently introduced conifers as profitable timber-producers in this country.

A comparison of the wood produced by the Weymouth pine in this country with that sent to the late Colonial and Indian Exhibition showed but slight differences, and nothing more than could naturally be expected between a mature and an only partially-developed tree.

The **Scotch Pine** (*P. sylvestris*) must on no account be omitted from our list ; as, especially for planting in exposed situations where few other trees could succeed, it is one of the best. Probably the principal reason why this tree has not been more generally cultivated, is on account of the low-priced timber it produces, for, of late years, it has been difficult, except in certain favoured districts, to get rid of it at a remunerative price. With the war, matters are quite different. The finest quality of Scotch pine timber, such as that produced in some of the northern Scottish counties, no doubt, is even now fairly remunerative; but, generally speaking, that grown throughout Southern Scotland, England as a whole, and also Ireland, is of so inferior a quality as hardly to fetch a remunerative price. No doubt, however, this pine will continue to be largely planted wherever shelter and ornament are of first importance: and rightly so, for few others are so well able to withstand cold, cutting blasts at high altitudes.

The **Douglas Fir** (*Pseudotsuga Douglasii*) is, in certain situations, a valuable timber-producing tree—indeed, as regards the actual production of timber in a given time, it is, perhaps, ahead of any other tree grown in this country. From measurements we have taken, the actual production of timber during fifty years was 240 ft., or nearly 5 ft. per year for half a century. In a plantation of the Douglas fir in Wales, planted twenty-two years, we found the average dimensions to be as follows: Height, 76 ft.; girth of stem at 24 ft., 4 ft.; cubic contents, fully 50 ft. The timber produced in this country is of excellent quality, being light but strong, works very readily, has a pleasant yellowish tinge, and takes on a good polish. That the Douglas fir is a tree that is eminently adapted for cultivation in this country is already well known; but to grow it in anything like a satisfactory way it must be planted in sheltered hollows, for extensive experience has long ago proved to us that it is ill-adapted for braving the storm, even at a few feet above the sea-level. Long ago we strongly advocated the forming of plantations of *P. Douglasii* alone, or with some other tree of about equally rapid growth, for, when mixed up with the general run of our forest trees, the leader, on overtopping those of its neighbours, soon gets broken over, or otherwise presents an almost branchless, whip-handle-like appearance. We do not wish to say one word against this our favourite fir: but the truth must be told, and our own experience, gained principally on a low-lying seaside estate, where the tree was annually planted by the thousand, is that *P. Douglasii* must occupy a sheltered situation if either ornament or utility be considered as points of first importance.

The **Giant Arborvitæ** (*Thuja gigantea*) is fast coming to the front as a British timber tree, and has already, at the hands of far-seeing planters, received a fair amount of attention. After a fair and impartial trial on our part, we have found it to be perfectly hardy, even at high altitudes, a fast grower and rapid timber-producer, a non-fastidious subject as regards the quality of soil in which it is planted, and one of the most easily managed and most accommodating of trees. The quality of timber produced in this country is such as to warrant us in speaking highly of it, it being of a desirable yellow colour, fine-grained, easily worked, remarkably durable, and light in proportion to its bulk. From the measurements of fully twenty-four specimens scattered over an English park, we have found that the average annual rate of growth is 22 in., but even this is exceeded by young trees.

The **Norway Spruce** (*Picea excelsa*).—Whether as a hardy, shelter-giving tree, or for the quantity of fairly good timber it produces, the common or Norway spruce must ever rank high in the list of useful trees that have been found suitable for culture in the British Isles. That it is a tree in every sense of the word admirably suited for extensive planting is acknowledged by all, as it luxuriates at high altitudes, and where fully exposed to our worst winds, and at the same time produces a great quantity of timber that has been found of excellent quality, well suited for general constructive purposes. As a shelter tree few others can equal the spruce, and when planted along the outskirts of exposed plantations the amount of warmth and protection it affords is quite surprising.

The **Silver Fir** (*Abies pectinata*) is another of those trees that have of late years fallen into disrepute, mainly owing to the increased importation of foreign timber. That it is an excellent and highly remunerative tree is unquestionable, and the very fact of its thriving luxuriantly on soils where the larch declines to grow should make it, in this country at least, of great value as a forest tree.

With the most satisfactory results has the timber been used for railway sleepers—in fact four sleepers laid experimentally have stood the wear and tear quite as well as those of Baltic timber, alongside of which they were placed. For roofing purposes the wood has likewise attained great fame, as it is found to stand vicissitudes of dry and damp alternately better than almost any other home-grown timber.

The coniferous trees just treated of are about the only kinds that can be recommended for profitable planting in this country.

The Atlantic Cedar and Japanese Larch might be added to the list, but present experience will not justify us in bringing any of these prominently to notice.

Twenty-five years ago, at the instigation of the then Earl of Derby, the writer formed several plantations on the Holwood Estate in Kent. At the outset it may be well to state that these plantations were not formed with the object of producing valuable timber, but rather for the ornamentation and privacy of the newly-acquired property. The trees used were the Scotch, Corsican, Austrian and Weymouth pines, Douglas fir, the larch, and several species of hardwoods. As all have succeeded well under exactly similar conditions, the following notes as to the rate of growth and production of timber, both of which are unusually great, during a period of twenty-five years may be instructive.

Previously to being planted the land, which may best be described as a hungry loam on a gravelly subsoil and sheltered, was let out for rough grazing and the cultivation of strawberries and other fruit.

The cost per acre of forming these plantations was :—

	£	*s.*	*d.*
Pitting, 2,722 at 1*s.* per 100	1	7	2
Planting	1	1	0
Trees, at 40*s.* per 1,000.	5	8	0
	£7	16	2

This price may appear both high and low, but in connexion with the former it should be explained that the coniferous trees when planted were about 16 in. high, the others about

3 ft., all being placed 4 ft. apart. Owing to the land having recently been under cultivation and labour at that time being cheap in the district, the opening of pits was carried out by contract at quite a nominal rate, the size of each being 12 in. square and 9 in. deep. After being planted the trees required little attention for the first six years, at which period they averaged 8 ft. in height, and the shade occasioned by the branch spread had killed out most of the grassy undergrowth.

As the plantations were primarily intended for ornament and shelter, the retention of the lower branches of the trees, at least along the margin, was imperative, and in order to ensure this, early thinning was engaged in and carried out at regular intervals up to the present time, always bearing in mind to allow the boundary trees plenty of room for branch development, those inwards, in order to induce clean growth, being left much closer on the ground. Though in the latter case the volume of timber produced is comparatively less than along the margins, yet it is of greater value owing to the trees being straight and clean-stemmed, the only exception being the Corsican pine which, even when isolated, has little inclination to form stout side branches.

The soil being light and resting on gravel was peculiarly suited for the growth of the pines, none of which suffered from disease or insect attack, though the Weymouth had occasional patches of the aphis with which it is usually attacked around London. The larch was practically free from canker.

During recent thinning operations a good opportunity was afforded of taking the actual measurements when felled of the various species of trees, these being as follows :—

Austrian	pine,	average	height,	46 ft. ;	cubic contents,	9 ft.
Corsican	,,	,,	,,	51 ft. ;	,, ,,	11 ft.
Scotch	,,	,,	,,	45 ft. ;	,, ,,	8 ft.
Weymouth	,,	,,	,,	42 ft. ;	,, ,,	6 ft.
Larch		,,	,,	47 ft. ;	,, ,,	8 ft.

It will thus be seen that the Corsican pine has surpassed all the others both in height and in the quantity of timber produced ; and in viewing the plantations from a distance the leaders of the Corsican pines soar quite 6 ft. above those of their neighbours. The Austrian comes next in the quantity of timber produced, but not in height ; and the larch and Scotch are of about equal size, the Weymouth being equal to the latter in height but not in bulk of stem. But the larch beats all in value of timber, for, while that of the various species of pine was difficult to sell at a remunerative figure, the larch wood was readily disposed of at a fair valuation.

My experience is that timber merchants fight shy of purchasing any of the pine family excepting the Scotch. This may be owing to prejudice or want of knowledge as to the value of timber produced by the Corsican and Weymouth ; but whatever the cause, the fact remains that the timber of both these species is difficult to dispose of at any but firewood rates. That of the Scotch being better known finds a ready market at about half the price of larch, which latter, after all, is the most useful and profitable of any coniferous tree cultivated in this country, and one for which the demand always exceeds the supply.

CHAPTER VII

PLANTING EXPOSED GROUND

As the majority of plantations for purely economic purposes are formed on the wind-swept moor or hillside, the successful management of these is a matter of considerable importance.

There are many difficulties to encounter in planting high-lying and exposed ground that one would never have to think about in low-lying and sheltered situations. Planting, for instance, should not be commenced until February or March, as, by deferring it till that time, the newly-inserted plants will be fresh and vigorous, and ready for an immediate start in growth, which would not be the case if they had been put out in autumn and subjected during winter to the inevitable wind shaking that is always to be reckoned with on exposed ground.

The choice of trees, too, for planting on wind-swept ground is a matter of far more moment than is generally supposed, for that there are certain species of trees peculiarly suitable for withstanding prolonged storms is well known to those who have had to do with the forming of woods and plantations at high altitudes. The size of trees planted has also much to do with the after success of the woods, and it may be well at the outset to say that these should not exceed about 12 in. in height on the most favoured sites to about from 6 in. to 9 in. on the more exposed grounds. They should also be properly prepared by frequent careful transplanting for the situations they are intended to occupy, as it can hardly be expected that a young and immature tree can, after being brought from a probably sheltered lowland nursery—as nine-tenths of those in this country are—to a high piece of ground, where at times the wind blows with terrific fury, and where one can scarcely stand upright, have sufficient stamina to stand against such odds and difference as must and do exist between the two points at present under consideration.

Sometimes it is well to trench over a piece of ground on partially-exposed land, and insert therein for a couple of years before the proposed plantation is to be formed, the young trees intended for planting out. This has a wonderful effect in hardening and inuring them to severe cold and a wind-swept situation. By forming a nursery of young trees by the sea-coast, I have known great success attend the formation of woods and plantations in maritime situations, and like results are sure to attend the planting out of trees in any uncongenial and unfavourable situation.

About the trees to be planted, being such as are sufficiently hardy to withstand prolonged storms, we will now say a few words. In the outer line—or, rather lines—the Scotch, Austrian, and Corsican Pines (*Pinus sylvestris*, *P. austriaca*, and *P. laricio*) are of first value, since they are able to stand against the storms of the hillside and produce a great amount of shelter to other less hardy kinds. It is often only necessary to make a wind-barrier, as it were—that is, to plant well-tried evergreen kinds, from which shelter can be obtained next the most windy position, then to follow up with other kinds that are second hardy in nature, and so on inwards ; while, in the very centre of the plantation, almost any kind of tree can be used.

The **Austrian Pine** has been successfully planted at high altitudes, and in the most exposed situations, and on the Continent it has proved itself of great value for breezy sites and maritime situations.

The State forests are composed of not a few Austrian pines, and they are greatly valued for the shelter and warmth they afford to other less hardy kinds of trees. The best results are to be obtained if planted when young, for when removed at a greater age, with roots large and rampant of growth, it is with great difficulty that they can be got to keep upright. Unfortunately the timber is of no special value.

The **Corsican Pine** is equally as good as the latter for using where winds are of common occurrence, proving stout and strong, rooting well, and presenting a broad surface of hardy evergreen foliage to the blast. It is a good timber-producer, and, being well fitted for growing in patches close together, will yet be largely used for forest work in this country.

About the **Scotch Pine** it is, perhaps, needless to speak, for every one who has travelled in Scotland, particularly in the more exposed northern parts, must have made himself acquainted with the capabilities of this valuable native tree. It can grow and flourish almost anywhere—on pure gravel, on the rocky mountain-slope, or by the rushing brookside, and in all these positions it seems to feel quite happy and contented, as the beautiful silvery glaucous foliage, the upright, rampant growth, and the cheery cinnamon or terra-cotta bark clearly indicate.

With these three excellent storm-resisting trees for an outer barrier almost any kind of planting can be engaged in, for the shelter they afford is amply sufficient to start away into rapid growth even our only second-class hardy kinds of trees. The **Sycamore** is a good tree for planting where the storms blow loud and long, being able to withstand in a very commendable way the first brunt of the hillside winds. It is also a good timber-producer, the wood at all times being easily disposed of, and at a very remunerative price.

The **Elder** and **Mountain Ash** are other valuable small-growing trees for planting on exposed ground, both flourishing apace even in very high and exposed woodlands.

In the **Scotch** or **Mountain Elm** (*Ulmus montana*) we have a first-class tree for planting as shelter, while the **Alder, Willows** of various kinds, and the **Hornbeam** should all receive attention in the formation of woodlands on exposed and storm-swept sites.

The **American Winged Elm** (*Ulmus alata*) has few equals for withstanding long-continued storms at high altitudes, for it sends out its cork-covered branches without fear of harm into the very teeth of the blast. I have noticed how well suited this elm is for planting on exposed, high-lying ground by the few examples that occur at considerable elevations in some of the screen-belts that have been formed on the flanks of the Snowdon range of hills, where the tree looks quite as healthy and happy as at lower elevations in a sheltered park.

The **Larch,** be it remembered, is a good tree for planting on exposed ground, for, though it gets twisted about and untidy of appearance, it has a wonderful recuperative nature, and will succeed well even when planted on the margins of exposed woodlands. By taking advantage of natural tree or shrub growth when forming plantations at high altitudes much good may be brought about. A young tree planted to the leeward of a clump of gorse, broom, juniper, or birch has a much better chance of succeeding than another planted where it has no shelter from the worst winds of the particular district. These wild clumps of natural shrubs should be encouraged in every way, for they will not only give a great amount of shelter, but help to thicken up the plantations as well. In exposed woodlands it is a good plan to plant up the margins with such hardy wild shrubs as the gorse, broom, thorn, juniper, blackthorn, etc., all of which will afford a great amount of shelter to the young plants when newly inserted, and help them to start away freely.

The **Common Beech** is a good all-round tree for planting in exposed sites, but especially where the soil is poor, or, in other words, composed principally of chalk or gravel. Some of the highest grounds in several of the English counties are occupied by thriving beech plantations, these acting as landmarks for many miles around, as notably at Knockholt and on the Chiltern Hills.

The **Oak** and **Ash** should both find places in high-lying and exposed woodlands, for, although they cannot be planted successfully along the margins, yet they thrive well where a little shelter is afforded, and where the soil is fairly deep and rich.

The **Birch** cannot be passed by in any list of trees for planting in exposed places ; it thrives well at high altitudes, and where only a small quantity of soil overlies the rock.

The **Wild** and **Bird Cherries** (*Cerasus vulgaris* and *C.*

Padus) are excellent ornamental trees for exposed grounds, where they grow to a large size and flower freely. They can both subsist where but a small depth of soil is present.

Preparation of the Ground and Planting.—As regards the pits for planting, these should be well prepared—that is, the soil loosened up and made free for the roots to run in, which will greatly help the trees to become quickly established—a point of great importance on exposed ground. The top turf may be chopped up and placed in the bottom of the pit, this serving, when decomposed, as manure to the roots, and assisting to promote vigorous growth. In planting, place the best-rooted sides of the trees to the windward or most exposed site, and do not cover with the soil to a greater depth than that in which the plant stood whilst in the nursery border. On very exposed sites, and where the soil is thin, notch planting and inserting with the planting iron are to be recommended.

It will be well, at stated intervals, to examine young plantations formed on high-lying and exposed ground, to see that the plants do not get shaken about with the wind and holes formed around the stems. This latter is highly injurious to the welfare of the plants, as the air passing down the stem side causes the roots to get dry to an inordinate degree.

Wind-swaying, where these holes have been formed around the stem, is also hurtful, as the tender roots get strained and barked, and ill fitted for maintaining a healthy condition of the young trees. The holes formed by swaying of the stems should be filled up with fine soil—not stones, as is sometimes the case—and a small piece of stiff, grassy turf tramped firmly against the stem on the side opposite that from which the worst winds may be expected.

Should rank vegetation, which, is, however, rarely met with at high altitudes and on exposed ground, interfere with the growth of the young trees, it would be well to have such cut over and either burned or spread evenly over the ground.

CHAPTER VIII

SEASIDE PLANTING

Few persons other than those who have actually been engaged in the work have the remotest idea of the difficulties to be encountered in the formation of belts and plantations on exposed and wind-swept seaside ground. To plant young trees around many parts of the coast of the British Isles, particularly where wide stretches of open seaboard are to be dealt with, without first erecting a shelter-screen of some kind or other, is useless work, and only productive of the most unsatisfactory results.

That there are not a few districts, however, where such a preliminary would be needless is well known, all that is required in certain instances being, first of all, to prepare the ground, and secondly, to suit the trees to the soil and situation, seeing that some varieties succeed better than others in certain soils and sites.

From experience I have found out how useless it is to plant in a haphazard way, at least, on the more exposed seaboards along the west coast, whereas, by careful manipulation, I have been successful in getting up shelter where before it was deemed almost an impossibility. The chief consideration in seaside planting is unquestionably shelter, be it only of a temporary kind, for it may be noticed anywhere along our coast that, wherever the direct force of the hurricane is broken, there trees and shrubs are growing best. Another evil—a great one, too—with which the planter has to contend is the injurious effect on trees, but more particularly on evergreen shrubs, of the saline particles which are driven and deposited with such force on the leaves and branches as in many instances to give them the appearance of having been scorched or cut over when in full vigour by an untimely frost. Wind-shaking, although a minor evil, must also be carefully guarded against, so that at the outset it is well to have the trees, except such as are of dwarf size, securely staked and tied, so as to obviate the dire results occasioned to the roots of newly-planted trees when the stems are allowed to rock to and fro with the wind.

Before commencing planting operations on the sandy and exposed sea-coast, the preliminary step is to erect a barrier of some kind, which will intercept the violence of the wind, and act as a screen or shelter to the young plants. For this purpose various kinds of erections are equally suitable, but that usually adopted, especially where loose sand alone is present, is a strong paling-fence thatched with brushwood. The posts should if possible be 6 ft. above ground and about 12 ft. apart, the paling-rails, two in number, being securely nailed to these at 2 ft. and 5 ft. from the ground; against this are placed spruce or gorse trimmings, these being 6 ft. long if possible, in an upright manner, and firmly tied to the cross-bars by means of binding wire or strong tarred rope. This may best be described as a dead fence, but it is, nevertheless, quite as valuable for the purpose required as a perfectly developed living hedge. By means of this a great advantage is gained and a favourable start for the newly-planted trees is secured. Then, compared with a stone wall, or, in fact, a wall of any kind, this screen-fence is greatly superior, the wind being broken up in passing through it, and, what is of as much value, also relieved of its saline particles, at least to a very considerable extent.

Where the shore is almost destitute of sand, and where certain kinds of vegetation subsist, I find it to be a capital plan to substitute for the dead fence just described that composed of turf and earth.

The raised mound should be not less than 5 ft. in height, and of sufficient width at top to allow of the planting of a double line of such plants as have been found most suitable for the wind-swept maritime situation in which they are to be used. For this purpose few plants equal, and certainly none surpass, the Sea Buckthorn (*Hippophœ rhamnoides*), Tamarisk (*Tamarix gallica*), the common elder, hazel, whitethorn, and at least three species of willow—*Salix Caprea*, *S. helix*, and *S. alba*. Immediately behind this screen, pits of not less than 2 ft. in diameter and about 18 in. in depth may be formed, and it will be all the better if some time is allowed to elapse before they are planted. The bottom and side of each pit should, where found necessary, be loosed with a pick, and if the soil is found to be of very inferior quality, it will be well, more especially where such can be readily obtained, to add a spadeful or two from some adjoining field.

Planting should not be commenced before March or April, an early start at growth being much in favour of young trees that have recently been transferred to the sea-coast. The plants to be used should not exceed, say, 12 to 15 in. in height, but of strong growth in proportion to their size, and supplied with an abundance of fibrous roots. Lanky, ill-grown, and coddled plants have but a poor chance of succeeding under such adverse circumstances.

In planting, keep the strongest roots seaward, and do not place the trees at a greater depth in the soil than that at which they stood whilst in the nursery border. The failure to use this precaution is a mistake, and is productive of anything but favourable results. Immediately behind this raised mound of turf, or the dead fence of branches above described, the best trees for withstanding the first brunt of the sea-breeze are the sycamore, elm, elder, and willow amongst hardwoods, and the Austrian and cluster pines (*Pinus austriaca* and *P. Pinaster*), to which might be added *P. laricio*, *P. sylvestris*, and *P. montana*, amongst conifers. These are all well-tried subjects, and may be relied upon as peculiarly suited for doing battle with hard-blowing and long-continued storms on the sea-coast.

Regarding the merits of the cluster pine for seaside planting, it would be almost superfluous for me to speak; while the elm and sycamore send out their stout branches into the very teeth of the blast, and are known as peculiarly well suited for such situations.

The following list includes such trees and shrubs as have, from long experience, been found well suited for planting on cold and wind-swept shores, and the trees are arranged according to their value in this respect.

Hardwoods

The **Sycamore** (*Acer Pseudo-platanus*) is, without doubt, the most valuable hardwooded tree of timber size that I have come across for planting in exposed seaside situations. It succeeds well, even at highwater mark, the stout, twiggy branches being thrown out into the very face of the blast. Even during winter, and in a leafless state, the amount of shelter afforded by this tree is quite surprising. Taking everything into consideration—its great hardihood, and suitability to various soils and situations, I consider the sycamore the most valuable of any deciduous tree that I have yet tried for seaside planting.

The **Elder** (*Sambucus nigra*) is, amongst small-growing trees, the best for planting in exposed seaside districts. Its powers of endurance are even greater than those of the sycamore, although the amount of shelter it affords is by no means so great. Where its branches are constantly exposed to the saline-laden breeze, and its roots amongst almost pure sand, it grows and thrives in a manner that is quite surprising. With the greatest advantages have I used the elder as a wind-break on some of the most exposed and desolate coasts of the British Isles. Even where grown as a single specimen, it seems to disregard the angry blast and saline particles with which it is almost constantly pelted—a fact that may be verified, in not one, but several stations along the coast. Then, what tree is of readier culture than the elder, succeeds better in poor, sandy soils, or spreads about to an equal extent?

The **Norway Maple** (*Acer platanoides*) stands the first brunt of the sea-breeze in a most commendable way—indeed, it may be recommended as one of the most hardy and valuable trees for the purpose under consideration that could be named. On the western coast of England, and in a very exposed and wind-swept district, I have used it with the greatest success in the formation of plantations, as it is of rapid growth and soon forms an excellent shelter to the other less hardy kinds of trees. It is by no means particular as to soil, but succeeds all the better if this is fairly rich.

The **Winged Elm** (*Ulmus alata*).—From a long experience of this, at present, little-known tree, I have every confidence in recommending it as one of the most valuable trees for planting in cold, wind-tortured, and maritime districts that have yet found their way into this country. Where many of our hardiest trees are bent and shrinking from the blast, this elm stands nobly out, and seems to defy both winds and storm.

The **Huntingdon Willow** (*Salix alba*) and **Bedford Willow** (*S. Russelliana*), but particularly the former, are excellent trees for windy shores. In many places along the coast the Huntingdon willow may be seen in a most flourishing and happy condition, and that, too, where the surroundings are the reverse of favourable. It is a tree of quick growth, and will succeed well in any class of soil if it be not too damp. In one instance that came under my notice, the trees were planted on a promontory overhanging the sea and in such a situation that they were almost constantly subjected to rough-blowing winds coming in from the Irish Sea, and yet they have grown with the greatest freedom, and to-day look as healthy and happy as if planted in some sheltered inland situation.

The **Beam Tree** (*Pyrus Aria*) is another excellent small-growing tree for planting in sites where, from cold saline blasts, few others could eke out even a miserable existence.

On the limestone cliffs of the Great Orme's Head this handsome and hardy tree grows in a most surprising way—indeed, with the exception of one or two species of willow, I question very much whether any other tree could exist under the trying circumstances. The hard and leathery leaves seem as if specially constructed for bearing storms, and, being plentifully produced, render a great amount of shelter to other trees.

The **Goat Willow** (*Salix Caprea*), for planting in almost pure sand on the sea-coast, is a most valuable small-growing tree, and it is so hardy that, without the least risk of harm, it may be planted at great altitudes, and in very exposed situations. In several of the maritime plantations that I have formed, this willow was used with remarkable success in the outer line or screen.

The **Aspen** (*Populus tremula*) is an excellent shore-tree, one that can withstand a great amount of rough usage, and as hardy and fast-growing a subject as could well be desired. In the formation of several exposed seaside woods on the west coast of England I planted the aspen largely, and, I may add, the results have been most satisfactory.

P. canadensis, **P. alba**, and **P. nigra** are likewise worthy of extended culture, for they succeed well by the seaside. They all stand the breeze from the sea, are perfectly hardy, and afford a great amount of shelter.

The **Mountain Ash** (*Pyrus Aucuparia*).—Although valueless, or nearly so, for the quantity of timber it produces, yet, as a hardy, free-growing tree, the mountain ash, or rowan tree, merits attention from planters of exposed seaside land. All along the coast of Great Britain this pretty tree may be found growing luxuriantly, and in such situations affording a great amount of shelter to other less hardy kinds of trees and shrubs. It may be planted without fear of harm, down even to highwater mark, and where the soil is of the poorest description. It is thus one of the most useful of trees for planting as shelter along the roughest and most wind-tortured parts of the coast.

The **Hoary Alder** (*Alnus incana*) and the **Common Alder** (*A. glutinosa*) can hardly be excelled for planting in wet portions of cold, wind-swept ground, and in the vicinity of the sea. Both trees grow with the greatest freedom, and are perfectly hardy, standing the first brunt of the saline blasts in a most commendable and praiseworthy manner.

I have found it to be a good plan where shelter is wanted on bare coast-tracts, if the quality of soil be at all suitable, to plant the alders pretty closely, and when they have attained to 15 ft. or 20 ft. in height, to cut every alternate one over at ground level. By so doing a number of stout shoots are thrown out early in the following spring, which, as they grow with great rapidity, soon fill up the blanks occasioned by cutting over the young trees.

After these have had several years' growth the remaining half of the original crop should be cut over, and when all have started from the stools a first-class shelter is obtained. Of course, damp portions of the ground should be selected on which to plant the alder.

The **English Maple** (*Acer campestre*) can well hold its own as a tree for planting on exposed seaside grounds. It is one of the hardiest trees with which I am acquainted, growing on exposed hillsides and at considerable elevations in a way that seems to attract the attention of every one.

Hedges or fringe fences of the native maple have succeeded amazingly in several maritime woods in which it was planted, and in places, too, where the wind blew loud and long.

The **Scotch Elm** (*Ulmus montana*) comes next on the list of such trees as I would recommend for the purpose under consideration. It affords plenty of shelter, as it grows freely from the root-stock, sending up numerous suckers, and is so hardy and proof against damage from storms that it may safely be planted in wind-swept districts by the seaside.

Birch (*Betula alba*) and **Ash** (*Fraxinus excelsior*) are other trees that repeated experiments have proved to be well suited for withstanding the ocean's blast and for planting in poor, rocky soils.

The **Turkey Oak** (*Quercus Cerris*), with just a small amount of shelter, will be found a most valuable tree for planting within the influence of the sea. It thrives well in very poor soils, and when in full leaf is capable of affording a great amount of shelter.

The **Evergreen Oak** (*Q. Ilex*) has proved itself to be

peculiarly fitted for planting as shelter in exposed and maritime districts. In the formation of seaside plantations, but particularly where, from the frequency and force of the storms, few trees can succeed, I have planted the evergreen oak with the greatest success.

The trees just treated of have no equals, among such as have yet been introduced, for withstanding cold seaside winds, a trial of many other varieties only resulting in proving their unfitness for planting in such situations. Two or three others, such as the **English Oak**, **Hornbeam**, **Beech** and **English Elm**, might be added to the list, but they can only be recommended for planting where at least partial shelter is afforded.

Coniferæ

Foremost amongst these I must, from long personal experience and a fair trial of several kinds under exactly similar conditions in every way, place the **Austrian Pine** (*Pinus austriaca*). It grows with unusual freedom, and affords a greater amount of shelter than any other tree with which I am at present acquainted.

That it will succeed equally well with the *P. Pinaster* when subjected to the sea-breeze, I am now fully convinced, while the amount of shelter it affords, and rapidity of growth, place it far ahead of that species for the purpose we are now considering. It may not succeed so well as the Pinaster when planted in pure sand, on the sea-coast, and this is the only point that can be adduced in favour of the latter species over *P. austriaca*.

In forming many large woods along the coast of Northern England I used the Austrian pine in preference to several others, and happily, too, for it has succeeded in a surprising manner, trees of ten years' growth being fully 13 ft. in height, and nearly as much in diameter of branch spread. With such a screen as that afforded by the hardy Austrian, many half-tender trees can be planted farther inland; and this is the method of procedure that I have found to be most successful in getting up shelter along bare and fully exposed parts of the shores of Western England and Scotland.

The **Cluster Pine** (*P. Pinaster*) and its smaller-growing variety *P. maritima* have a world-wide reputation for their suitability for planting on exposed seaside tracts of ground. That they are of great value for planting on sandy wastes is a fact that cannot be gainsaid—indeed, few other trees could succeed or eke out an existence in pure sand and where the roots come in contact with the salt water. A great drawback to this pine is its long tap-root and want of fibrous roots; it transplants with difficulty, but this, as in various other species of pine, may be greatly obviated by careful nursery management.

In my own opinion the typical tree is hardly equal to the variety *maritima* either for shelter or withstanding the sea-breeze. The variety, too, is, if anything, the most valuable for shelter-giving, it having a much greater inclination to retain the lower branches intact.

The **Corsican Pine** (*P. laricio*) quite equals the Austrian in its powers of withstanding long-continued and cold winds. That it does not succeed so well on the sea-coast is a fact of which, from repeated experiments, I am fully aware. The Corsican pine, too, is a valuable timber producer—a fact that is well worthy of consideration in extensive planting.

In the **Giant Arborvitæ** (*Thuja gigantea*) we have another excellent addition to the list of trees that have been found suitable for planting on exposed maritime grounds. It grows with great rapidity, and I have never found even a solitary example of this tree having been uprooted or injured during the most severe storms. On the sea-coast of Wales I have used the giant arborvitæ largely in the formation of woods and plantations, and with great success. It transplants well, even when of large size, and is readily propagated.

Pinus montana may also be recommended for afforesting tracts of ground by the sea-coast. It is a tree of undoubted hardihood, withstanding cold and cutting winds in a worthy manner.

The **Scotch Pine** (*P. sylvestris*), though by no means equal to the above for planting by the seaside, must on no account be omitted from our list, for it is a hardy, fast-growing specimen, and one that can do battle with very severe and long-lasting storms. It should not be planted where it will meet the first brunt of the storm, but given a little shelter, such as that afforded by the above-named kinds. The varieties of pine just named may all be relied upon as peculiarly well fitted for the purpose under consideration.

The **Large-fruited Cypress** (*Cupressus macrocarpa*) and **Pinus insignis** might be added to the list, both being well suited for maritime planting—inland, however, they cannot be relied upon. *Pinus insignis* grows with unabated vigour on the wind-swept coast of the Isle of Anglesey, and stands the breeze in a most surprising and praiseworthy manner.

The **Atlantic Cedar** (*Cedrus atlantica*) and **Swiss Stone Pine** (*Pinus Cembra*) do fairly well as seaside trees.

Shrubs

The **Sea Buckthorn** (*Hippophea rhamnoides*) is unquestionably the best all-round shrub for planting as a shelter by the sea-coast with which we are acquainted. To see it in several districts of Scotland, growing amongst almost pure sand, and where constantly exposed to fierce blasts, shows how valuable a shrub it is for exposed maritime situations. For affording shelter it is one of the best shrubs that I know of, the unusually twiggy branches sifting and dividing up the wind in a most remarkable manner.

Tamarix gallica and **T. germanica** are two excellent seaside shrubs, and such as are well suited for planting on exposed places. They grow with great freedom, transplant well, and are readily propagated. Even in pure sand they seem quite at home, growing freely, and producing a rich abundance of their showy flowers. In very exposed parts of the coast, and when fully open to direct sea-blasts, we have frequently seen the tamarisk fully a dozen feet in height, and nearly as much in branch-spread. Though little known. *T. germanica* is a most valuable seaside shrub, about 6 ft. in height, with small leaves, and spikes of conspicuous red flowers.

The **Box Thorn** (*Lycium europæum*) may be introduced without fear of harm to the seaside woods, for it is not only perfectly hardy in such situations, but it bears constant exposure to wind as well as any shrub I know of. Hedges of this plant have been formed in many maritime parts of England and Wales, the only support being a few stakes driven in here and there along the line of fence. It will grow in pure sand, and when lashed by the waves.

The **Snowberry** (*Symphoricarpus racemosus*) comes next on the list of shrubs that have been found suitable for the purpose under consideration. It is a plant of extraordinary hardihood, one that increases rapidly from the rootstock and affords a fair amount of shelter.

The **Tree Mallow** (*Lavatera arborea*) can hardly be surpassed for shore planting, where it frequently attains to a height of 10 ft. It has been found of great utility in some of the islands along the coast of Scotland.

Spirea Adiantifolia is a fitting companion for the latter, growing and flowering very freely in wind-swept gardens along the Scottish coast.

Griselinia littoralis has stood a severe test as to its capability for withstanding cold winds blowing in from the sea.

The **Dogwood** (*Cornus sanguinea*) and **Flowering Currant** (*Ribes sanguineum*) are, likewise, excellent seaside shrubs, of perfect hardihood and readily propagated.

The **Tree Purslane** (*Atriplex halimus*) is also a really valuable shrub that is totally indifferent to the salt spray, and from its dwarf, evergreen nature, and silvery-scaly leaves, is also more or less ornamental. It does not seem to mind either wind blowing direct from the sea, or whether the soil in which it is planted is of only moderate quality.

The **Laurustinus** (*Viburnum tinus*).—Where the situation is not too exposed, this is a most valuable and ornamental shrub.

Lilac of various species are well adapted for planting by the seaside, the two kinds most to be recommended being the **Common** and **Persian** (*Syringa vulgaris* and *S. Persica*).

The **Spanish Broom** (*Spartium junceum*) has been used with the greatest success all along the coast, being perfectly hardy, and an excellent subject for cold, draughty positions.

The **Tree Groundsel** (*Baccharis halimifolia*) is not only a shrub of great beauty, but one that is perfectly hardy, and well adapted for planting by the seaside. It will thrive in almost pure sand, but it is all the better for a poor gravelly loam, and seems to delight in the ozone of the seaside atmosphere.

Both the **Portugal Laurel** (*Prunus lusitanica*) and **Sweet-Bay** (*Laurus nobilis*) are valuable evergreen shrubs for seaside planting. From their large size and compact habit, they afford a great amount of shelter.

The **Common Holly** (*Ilex aquifolium*) and its golden form both do well at the seaside, and this may also be said of the **Common** and **Scotch Laburnums**. They may be used with best advantage where the direct force of the blast is broken up.

The **Common Gorse** (*Ulex europæus*) and the **Broom** (*Cytisus scoparius*) should on no account be omitted from our list of shrubs that are valuable for maritime wastes where the wind exerts its full power.

The above shrubs include the principal of such as can be recommended for planting along the sea-coast, but where good shelter is afforded by the trees named in this chapter, a few others might be recommended for trial. These include the **Strawberry Tree** (*Arbutus unedo*), **Euonymus japonicus**, **Berberis Darwinii**, **Ligustrum Ovalifolium**, **Daphne Mezereon**, and **D. laureola**, **Ruscus aculeatus**, **Hypericum calycinum**, **Vinca major** and **V. minor**, and several kinds of thorn.

The **Maram**, or **Sea Matweed** (*Psamma arenaria*) is one of the most useful grasses with which I am acquainted for binding shifting sands on the sea-coast. Not only so, but by using it as a pioneer, the amount of shelter it affords renders other more difficult subjects by no means hard to cultivate. It usually attains to a height of from 2 ft. to 2½ ft., much depending on the situation, whether partially sheltered or fully exposed. The root-stock creeps widely, some specimens that have been followed up in the sand being of the amazing length of 35 yards. Amongst loose and drifting sand the running roots find what is most suitable for the welfare of the plant, and it is astonishing with what persistency they bind in an unusually short space of time these shifting hills of almost dust-dry sand. In planting, place the plants in parallel lines, about 16 in. apart, and at a distance of 12 in. from each other. Large plants may be subdivided to almost any extent. A garden line is stretched along the ground, a notch 10 in. deep taken out, the grass inserted therein and filled with sand, and afterwards firmly trodden. That the sea matweed is a most useful plant for fast gaining a footing on sandy tracts of sea-coast, and thus allowing of following up with the shelter-giving trees, cannot be disputed.

The **Lymegrass** (*Elymus arenarius*) is also valuable for growing in almost pure sand on the sea-coast.

It is of tall, elegant growth, and is readily increased from root divisions.

The following trees and shrubs, alphabetically arranged, are recommended for seaside planting :—

Trees for the Sea-coast.

Acer campestre
——— creticum
——— Pseudo-platanus
Alnus glutinosa
——— incana
Betula alba
Carpinus betulus
Cedrus atlantica
Cupressus macrocarpa
Fagus sylvatica
Fraxinus excelsior
Pinus austriaca
——— laricio
——— montana
——— Pinaster
Pinus Pinaster maritima
Populus canadensis.
——— nigra
Pyrus Aucuparia
——— Aria
Quercus Ilex
——— Robur
Salix alba
——— Caprea
——— Forsteriana
——— Russelliana
Thuja gigantea
Ulmus alata
——— campestris
——— montana

Shrubs for the Sea-coast.

Atriplex halimus
Aucuba japonica
Cerasus lusitanica
——— Padus
Cytisus Laburnum
——— scoparius
Euonymus japonicus
—— europæus
Fabiana imbricata
Griselinia littoralis
Hippophæ rhamnoides
Ilex Aquifolium
——— Aquifolium aurea
Laurus nobilis
Lavatera arborea
Lycium europæum
Rhamnus frangula
Ribes sanguineum
Rosa spinosissima
Shepherdia argentea
Spirea adiantifolia
Syringa persica
——— vulgaris
Symphoricarpus racemosus
Tamarix gallica
——— germanica
Ulex Europæus
Viburnum tinus

Grasses.

Elymus arenarius
Psamma arenaria

CHAPTER IX

TOWN PLANTING, AND THE TREES AND SHRUBS THAT ARE BEST ADAPTED FOR SMOKY LOCALITIES

PROBABLY no work connected with horticulture requires more judgment and good management than the planting of trees and shrubs in urban districts. The materials and soil of which streets and town gardens are usually formed are ill fitted for maintaining a healthy condition in trees and shrubs for any length of time. This fact, coupled with the impurities of the atmosphere in densely populated centres, has to be constantly borne in mind. In more favourable districts all that is necessary is to open a pit of sufficient size to contain the roots of the tree or shrub to be planted; but in towns the soil, often as hard as iron and composed mainly of refuse building materials, contains but little plant food. For many years past careful observations have been made, not only in London, but in Glasgow, Liverpool, Manchester, Warrington and Dublin, as to which trees and shrubs succeed best in the smoky localities of each town, and it is mainly by tabulating these different experiences that satisfactory information on the subject has been obtained. Coal smoke from the chimneys in the larger and more crowded centres of industry is no doubt bad enough, but, when we have to contend with an atmosphere that is largely impregnated with the outcome from chemical, gas, or iron works, the difficulties to be encountered are correspondingly increased.

The injurious effects of smoke have become much more pronounced during the past century, and Sir William Richmond, R.A., told the annual meeting of the Coal Smoke

Abatement Society that Westminster Abbey had suffered from more rapid decay in the last hundred years than in all the previous centuries of its existence. The chief cause of the destruction of the stonework has been shown to be the presence in the air of sulphur acids : the stone is converted into sulphate of lime ; in the process of its formation this disintegrates the stone by expansion. The connexion between smoke and stone decay appears to lie in the action of invisible gases emitted from the smoke particles.

If stonework suffers so at the hands of smoke and sulphuric and other acids, what, it may be asked, must the effect be on the foliage of trees and shrubs—particularly such as are planted in the most smoke-infested parts of our great towns and cities ? When compared with Continental cities—Paris, Brussels or Berlin—where tree culture is carried out most successfully, the atmosphere of British towns is impregnated to a far greater extent with noxious fumes. Dry low-lying and confined areas, particularly where excessive heat and atmospheric impurities are present, are decidedly the worst, while open and high-lying districts, though in the centre of a town, offer fewer difficulties.

That certain trees and shrubs succeed best in particular towns is a well known fact, and the smoke-proof London Plane is by no means the best tree for some of the colliery districts ; in Sheffield, for instance, its place is largely taken by the Canadian Poplar. In Manchester, the Lime would appear to thrive best, after which the Elder, Thorn and Plane succeed in the order named. The variegated-leaved Sycamore and the horse chestnut are favourites where the smoke from collieries is most offensive. But many such cases could be pointed out, and even in the case of bedding plants certain species succeed best in particular localities. In the gardens about the Royal Mint, where they are exposed to the deleterious fumes from gold-refining works, Fuchsias do remarkably well ; indeed, the dwarf edging variety, Golden Treasure, thrives so well that advantage has been taken of the fact to propagate some of the stock that is annually required for one of the London parks from cuttings taken at the Mint. In the East End of London the Creeping Jenny (*Lysimachia*) thrives well as a window plant, while in the chemically impure atmosphere of Lambeth one of the Veronicas is the favourite plant for indoor culture. The St. John's Worts (*Hypericum*) do not as a rule thrive well in London ; yet around the Tate Gallery, which is only divided by the river from the Lambeth pottery district—the worst in the metropolis for atmospheric impurities—one species at least flourishes amazingly, and has produced flowers in abundance for many years past ; while at St. Paul's Churchyard, the lesser Periwinkle (*Vinca minor*) has become quite established and runs about freely. In Chancery Lane, at the Record Office, the common Ivy, Bladder Senna, and Yucca do best. In other parts of London, too, well known varieties of Campanula are largely grown as pot plants. It is a somewhat strange fact, too, that some varieties of trees and shrubs succeed better than the type species in smoky localities, as witness the London Plane (a variety of *Platanus orientalis*), variegated-leaved Sycamore, fastigiate Poplar, two varieties of Pyrus, Weeping Elm, Weeping Ash, and several varieties of Acacia, notably *Robinia pseudo-acacia inermis* and *R. pseudo-acacia Bessoniana*.

Similarly, amongst shrubs, we have the dwarf Holly, golden variegated Euonymus, golden Privet, Ribes, double-flowered Gorse, *Euonymus radicans variegata*, and others. With Grasses, too, some curious experiences might be related. At the British Museum the Yarrow completely ousted the Grasses from the plots in front of that building, and in the moat of the Tower of London several Grasses that succeed in less smoky parts of the metropolis quickly die out. Near the main entrance to the Tower of London, and close to Billingsgate Fish Market, considerable difficulty was experienced in getting the Plane trees established ; though in the matter of soil, and choice of strong sturdy specimens, every possible care was taken. At last it was found that the drip from the fish carts was the cause of the evil, and a remedy was quickly found. In another garden, where dust, smoke, and soot are plentiful, the Bladder Campion (*Silene inflata*), *Saponaria officinalis*, the common Marigold, and Rye Grass seem positively to revel. In situations almost constantly subjected to the sulphurous fumes of the railway engines near Camden Town, and in the poorest of soils, *Poa annua* would appear to be quite at home. The chemical fumes from the pottery works at Lambeth are well known to act injuriously on vegetation generally, but the Mulberry, Fig, Sycamore, Turkey and Evergreen Oaks thrive as well there as they do in any part of the metropolis. The fumes given off from many of our city manufactories act most perniciously on vegetation generally—a fact that was brought to my notice by the behaviour of some of our most valuable smoke-resisting trees and shrubs planted in the graveyard at St. Giles-in-the-Fields. Meeting the gardener there I remarked on the wretched condition of the trees and shrubs generally, his quick reply being, "Well! with Crosse & Blackwell's on the one side, and Nixey's Black Lead Works on the other, it's a wonder there's a living plant left!" Here the common Fig and Black Poplar seemed better able to withstand the atmospheric conditions than either the London Plane or Acacia. With the largely increased use of coal gas for cooking purposes, improved grates, and the better combustion of fuel, the atmosphere of certain districts of London, has, however, become much less smoky than was the case a few years ago, and in consequence vegetation generally succeeds better. This is especially the case in some of the low-lying districts adjoining the Thames where the "slot" system of providing gas for cooking purposes has caused a great decrease in the consumption of coke and coal, with a corresponding reduction of the attending evils of smoke and soot. In one of the poorest parishes many of the inhabitants have taken advantage of the facilities offered by the Gas Company in the matter of cooking by means of gas provided by the "slot" system, which, in comparison with coal, has been found not only cheaper, but cleaner and handier to use. According to a competent authority the smoke nuisance has, in consequence, greatly abated, and with the purer air, the cultivation of window and other plants, as well as of trees and shrubs, has been to some extent simplified. The electrification of the Underground Railway has also had a beneficial effect on vegetation. Although we cannot prevent fog, which is an atmospheric condition, yet much can be done to prevent it being a dirty fog, and during the past five years much has been done in that direction. Better roads with less dust also assist largely in keeping the atmosphere of London in a pure condition.

With the rage for coniferous trees which was at its height about half a century ago, it is not surprising that several species of Cypress and Cedar, the stately Pines and Arbor vitæs, as also the Araucaria and Junipers, found their way into our town gardens and squares. Hosts of evergreens, too, from almost every part of the world were introduced into London, but few have been able to survive the smoky and otherwise impure atmosphere of the great metropolis.

Deciduous trees and shrubs, both flowering and ornamental leaved, should certainly be regarded as the *sine qua non* of the London planter. Amongst evergreen trees few are suitable for town planting, and, though a limited number of evergreen shrubs may succeed for a time, yet the list of deciduous species is far more extensive. We have only to take notice of such evergreens as the Holly, Rhododendron, Laurel and all the conifers, with probably one exception, to find how useless it is to plant them in expectation that they will give satisfaction. This is not hard to account for, as in winter, when the fires are all alight and smoke and soot the order of the day, the leaves of evergreens are fully developed and in the best possible condition for reaping the attending disadvantages. With deciduous species the case is quite different, for these are, so to speak, asleep when the deadly smoke and vapours are most abundant in our towns and cities. I am quite aware that one occasionally sees evergreen shrubs and trees in a fairly thriving condition; but it should be remembered that in the majority of such cases they were planted when conditions were much more favourable than at present.

It is, perhaps, to be regretted that evergreen shrubs

do not succeed better in smoky localities, being planted principally for their refreshing colour in winter ; but if our atmospheric conditions utterly preclude the use of such, then it is folly to throw away money on useless planting, and the winter aspect of deciduous trees and shrubs is infinitely preferable to that of unhealthy evergreens. The bursting into leaf of the deciduous tree or shrub is not shared to a like extent by evergreens, which lack that delightful changeableness and the interest that is attached to spring growth.

It is difficult to define accurately the boundaries of a town or the worst smoke-infested areas, as far as tree and shrub growth is concerned. In London, for instance, certain trees and shrubs which positively refuse to live in the heart of the city do fairly well in the suburbs, while still further out, where the atmosphere is comparatively pure, they may thrive in quite a satisfactory way. These thriving and non-thriving areas are sometimes very sharply defined, and this has given rise to a false idea regarding certain trees and shrubs that will rarely succeed in the more smoky parts when compared with the same species which are found to do well in the outer suburbs. High-lying and fairly open parts of a town are also far more conducive to plant growth generally than those that are close and confined. In town planting there is, however, no necessity for the almost monotonous repetition of such trees as the Plane and Lime, or amongst shrubs, of the Privet and Lilac, for there are many others that will do almost equally well, and that are quite as ornamental. Probably the fact that such are not well known would form an excuse for their absence, and it is to be hoped that at least one object to be accomplished by the writing of this book will be a greater interest in, and wider knowledge of, the various species of trees, shrubs, and plants generally that from long experience have been found suitable for planting in the town garden.

Preparation of the Ground.—Generally speaking, the materials with which roadways are made are not only unsuitable for tree cultivation, but positively destructive to vegetation of almost every description. This also applies to our squares, terraces, and open grounds around houses, the soil of which is little other than refuse building materials, and mainly composed of broken bricks and stones, gravel, old mortar, iron, wood and shavings. In such a medium it is perfectly useless to look for that healthy and vigorous growth which is so essential in street trees, that are still further handicapped by having to do battle above ground with the impurities of a town atmosphere.

Having for a number of years had to plant trees and shrubs in many parts of London, it was found that in nearly every instance substituting good soil for that found naturally was a first necessity. Many failures in street planting from this neglect of providing suitable soil could be pointed out, the result being that a section of the public has become tired of the subject in consequence of the initial expense and subsequent failures. Too often, also, the important operations of preparing the ground and planting the trees are left in the hands of the surveyor or builder, who has little or no knowledge to fit him for the work, the operation being carried out by labourers who are also entirely ignorant of what is required or the conditions necessary for successful tree and shrub culture. The consequence is that failure is almost certain, and the trees which were strong and healthy when planted gradually become unhealthy and ultimately succumb to a combination of circumstances which were brought about by the ignorance of the operators.

Another source of failure in street planting is the generally pent-up condition of the roots, for in several cases that have come under our notice lately the planter seemed to think that it was quite enough to cut a small hole in the pavement or street of sufficient size to hold the roots of the tree to be inserted. Were the soil free, as we find in a field, this system might answer; but where the roadway is hard as iron and composed mainly of clinkers and gravel, the case is totally different. Another fruitful source of failure in street planting is placing the pavement in too close proximity to the stem of the tree, and numerous instances could be pointed out where even old and established specimens have suffered irreparable damage in consequence of having the paving brought up too close to the stems.

The roots should always be allowed plenty of breathing room, and to affect this a good-sized space should be railed off around each tree and no pavement laid within it. Gratings may be placed on the surface of the ground around the tree, should circumstances compel such a course. By adopting either plan, a double benefit to the trees is brought about by allowing free access of water to the roots and preventing an accumulation of noxious gases in the soil, as would ensue if the flagstone or pavements were used.

Where street trees are to be planted, the ground surface should in every instance be thoroughly broken up for a space of not less than 8 ft. square, and to a depth of, say, 4 ft., the inferior soil removed and replaced by that of good quality, preferably of a loamy description, or loam and leaf soil in about equal proportions. Before placing the fresh soil in position, the sides and bottom of each pit should be thoroughly loosened with a pick or fork. By undermining the sides of each pit, a much larger and freer root run will be provided, and this will not necessitate so much of the street or pavement being torn up as if the pits were of equal width at the top and bottom. We have found, in London at least, that the addition of a small quantity of leaf soil to the loam is highly beneficial to the growth of trees by retaining dampness and encouraging root spread. The newly added soil should be firmly trampled in the pit before planting is engaged in. Sometimes, where the original soil is not of too inferior quality, a small proportion has been mixed with the loam and leaf soil, but, speaking generally, this course cannot be adopted.

In squares and gardens where shrub-planting is to be engaged in, a general renovation of the soil is also imperative, and this can best be done by thoroughly trenching the soil to a depth of, say, 4 ft. and adding a large proportion of fresh loam or other soil. Deep trenching and thoroughly breaking up and loosening the soil is a most important factor in town planting, and should never be neglected.

Manure is sometimes added to the soil, but it is objectionable from several points of view, and, if used at all, should be thoroughly decomposed and incorporated in small quantity. Good loam and leaf soil is infinitely preferable, and, where necessary, sandy soil makes a good addition. A little fresh lime added to the soil has been found most beneficial in town planting, and in old and exhausted borders, where the soil has become tainted with chemical impurities, the value of lime or chalk as a cleansing agent is not sufficiently appreciated. Of course, where so-called American shrubs are to be planted—which is, however, rarely attempted in London—neither chalk nor lime should be added to the soil. This question of soil is so important that no one planting street trees or shrubs can afford to ignore it, and while the extra cost in providing it is but little, the advantages gained are great.

Preparing the Trees and Planting.—Trees intended for planting in towns, and especially alongside streets and footpaths, should be specially prepared in the matter of transplanting and pruning. As tree guards are a necessity in protecting trees by the sides of streets, the trees must have their stems free of branches ; therefore the buds and branches on the stems need to be removed for a distance of 6 or 7 ft., whilst surplus leading shoots and ungainly branches should at the same time receive attention in pruning. The tree also needs to be frequently transplanted in order that an abundance of fibrous roots may be produced, and every effort made to produce healthy, vigorous specimens suitable for the uncongenial surroundings of their permanent quarters. In street planting it is advisable to use trees that are from 12 to 14 ft. in height ; and if these, for some years previously to their final shift, have been specially prepared in the way of frequent transplanting and careful pruning, little fear for their future welfare need be entertained. Autumn or early spring planting is to be recommended, the former time being in most cases preferable. As little time as possible should be allowed to elapse between the lifting of the tree in the nursery and its being transplanted in the new position. Spread the roots out to their full extent around the stem and avoid planting too deeply; the nursery mark on the stem serves as the best guide as to the depth at which to plant. Planting too deeply under the mistaken idea that doing so will secure the tree in the ground is a fruitful source of decay and ultimate death of many street trees planted in the metropolis, and it is not uncommon to see whole avenues of trees that have made little or no progress for many years owing to this cause. After a tree has been placed in an upright position on the prepared site and the roots properly disposed, the soil should be filled in and trampled firmly both amongst and over the roots. It may be well to warn planters against the pernicious practice of allowing leaves, packing materials, or grassy turf to come in contact with the roots of newly-planted trees. In dry situations a saucer-shaped hollow may be left around the stem of the newly-planted tree, while mulching applied during dry and warm summers is to be recommended. The planting of shrubs should be carried out with as much care as is used in the case of trees. It is preferable to trench land in which shrubs are to be planted rather than to make a separate pit for each shrub.

Street trees should be carefully matched—that is, those of similar height and shape used in the same street. Too often this principle is not observed, as in a new street near the Strand, where some of the specimens are about 10 ft. high, others, 15 ft. and 20 ft. high—forming a very irregular, badly matched row. Many newly-planted town trees are destitute of leading shoots and have ungainly side branches—faults that should never be permitted when choosing specimens for such an important purpose. In choosing trees for street planting, the following rules should be observed :—

1. Stout, healthy, well-rooted and recently transplanted trees should alone be chosen.

2. They should, for the same street, be of nearly equal height and branch spread.

3. Straight-stemmed trees, with stout leading shoots, are to be preferred.

4. The height should range from, say, 12 ft. to 14 ft. or

over, and the strength of stem should be proportionate to the height.

5. Trees with wand-like, crooked, or cankered stems should be avoided in street planting.

6. They should be beautiful, shade-giving, and easy of culture.

Fencing and Staking.—In order to prevent damage, newly-planted trees should be fenced and staked at once. Of fences or guards there are many kinds: they are made of wood, wire or iron. The iron tree-guard has many advantages over those of wood or expanded metal, and, being made in two sections, it can be readily placed in position after the tree has been planted. For trees from 12 to 14 ft. high the guards need not exceed, say, 7 ft. in height, and preference should be given to those in which the uprights are bent outwards at the top; for this not only lessens the risk of interference with the branches, but is pleasing in appearance. Sometimes it may not be considered necessary to protect town trees, particularly such as are growing in side streets, or in squares, but in every case firm staking is necessary in order to prevent damage from wind.

Wooden tree-guards consist of about half a dozen poles or uprights, about 7 ft. long, joined together around the tree trunk by means of wire. When compared with those of iron, they have, however, several disadvantages; for they may be climbed with ease and they do not last long. Where it is found sufficient to stake the trees without having recourse to guards, ash poles, from 2 in. to 3 in. in diameter, and 10 ft. high, should be driven firmly into the ground as close to the stem as possible. The tree should be tied with specially prepared tar rope, which should be crossed between the stem and stake to prevent damage by chafing. From time to time it will be necessary to see that the band of string does not become too tight. On rare occasions only is it necessary to stake shrubs, but this is sometimes advisable in exposed positions or in the case of shrubs of unusually large size.

Watering and after Management.—For a few years

after being planted, trees and shrubs will require a certain amount of attention in the matter of watering and mulching during the prolonged heat and drought, and also in preventing the evils attending excessive wind-swaying owing to the stakes and moorings becoming defective.

The goat and wood leopard moth, as also numerous kinds of caterpillars, attack newly-planted town trees, particularly the various species of Pyrus, Thorn, Willow and Poplar. The former tunnel into the main stem and render it so weak at the point attacked that it readily breaks across during windy weather, while the caterpillars feed on the foliage and greatly impair the health of the trees attacked. Spraying, hand-picking and shaking are to be recommended in the case of the caterpillars, while, to combat the goat and leopard moth, a small quantity of cyanide of potassium can be inserted in the mouth of the tunnel in order to dislodge or kill the insect.

Watering should preferably be performed in the evening, the ground around the roots being thoroughly soaked, while mulching with freshly-mown grass or old straw will prevent the too rapid evaporation of the moisture. In extreme cases binding the stem with a hay rope is to be recommended. In order to prevent straining of the roots and bark chafing, the moorings of trees should receive a periodical examination.

The following trees and shrubs are to be recommended for town planting :—

Trees

The **Oriental** or **Common London Plane** (*Platanus orientalis acerifolia*).—This variety of the oriental plane stands first in the category of select town trees. Not only does it grow vigorously in towns, but it is peculiarly well adapted for withstanding smoke and other impurities of the atmosphere. Repeated experiments have clearly proved that in London this tree flourishes better than any other, and a visit to the Thames Embankment and several of the urban districts will substantiate the statement ; while the fine old tree which still exists in Cheapside, and the equally beautiful specimen which has hardly room for perfect development in the Court of Stationers' Hall, Ludgate

PLANE TREE IN REGENT'S PARK

Hill, afford other examples of how well suited this handsome tree is for doing battle with the adverse conditions peculiar to the great metropolis. As a diversity of opinion has existed about which variety of plane it is that grows with such vigour in and around London, it may be stated that, on a careful examination of a large number of specimens, the variety *P.O. acerifolia* was found not only more commonly distributed, but, likewise, better suited for town planting than the typical *P. orientalis*. This valuable variety is readily distinguished from the normal plant by the less deeply divided leaves, and from the American plane (*P. occidentalis*), with which it is sometimes confounded, by the many fruit "balls" which are attached to each peduncle, the fertile catkins of *P. occidentalis* being, for the greater part, produced singly.

But not only for its value as a town tree is the oriental plane much sought after, but the giant proportions to which it attains, coupled with the handsome, finely-cut leaves and easy habit of growth, render it one of our most desirable ornamental trees. Further, it is of the easiest culture, succeeding extremely well in soils of very opposite qualities, Taking everything into consideration, we question whether any other of our forest trees is equal in value to the plane for purposes of town planting.

The **Maidenhair Tree** (*Ginkgo biloba*).—The maidenhair or ginkgo tree is one of the most valuable that can be planted in the impure atmosphere of a town garden. Few trees can compare with the one in question for withstanding the deleterious effects produced on vegetation generally by coming in too close contact with the impurities of our great centres of industry. The ample delicate-green foliage betrays—even late in the season, and when about to be cast off—little evidence of the fierce struggle that must almost constantly go on between vegetation and the smoke and filth of our towns and cities. That the thick, leathery leaves and strong constitution of the tree play an important part in keeping it free from disease is clearly evident, while the annual renewal of the leaves enables it to cast off the sooty nodules which work such havoc on the tender foliage of most evergreen trees.

At several places in and around the great metropolis—and in places, too, where one is almost stifled with the fumes from chimneys—this tree may be seen in almost as fresh a condition as those in the open country.

The **Ailanthus** or **Tree of Heaven** (*Ailanthus glandulosa*) may be seen in a very flourishing condition in many of the largest centres of industry in Southern England. It has been largely planted in many Continental cities, and has proved itself one of the few trees that is capable of withstanding the impurities of a town atmosphere.

The **Black Italian Poplar** (*Populus monilifera*).—Next to the plane amongst forest trees I consider the black Italian Poplar to be the most valuable for planting in smoky towns. As a proof of this there are to be seen numerous fine specimens of this tree in a flourishing condition, and clothed with the most healthy foliage, in some of our large cities—to wit, London, Glasgow, Liverpool and Manchester.

The **Canadian Poplar** (*P. canadensis*), and its variety, *P.C. nova*, are excellent trees for planting in smoky localities. The former succeeds admirably in the very centre of Sheffield, in the old parish churchyard, where for hundreds of yards around not a particle of living vegetation is to be seen. The variety *nova* is a very superior tree for street planting, it being far more ornamental, and of more rapid growth than the black Italian poplar, and equally noticeable for retaining a healthy and flourishing condition under the adverse circumstances connected with a town atmosphere.

The **Abele Poplar** (*P. alba*) grows with great freedom when subjected to smoke and foul air. In the very heart of our largest towns it may be seen flourishing in a manner that is almost incredible.

The **Lombardy Poplar** (*P. fastigiata*) is another tree that has been planted with some success in and around many of our largest cities, but it cannot equal any of the foregoing for withstanding the baneful effects of a tainted atmosphere. In the outskirts of towns, where air is purer than amid chimneys pouring forth their volumes of smoke, the Lombardy poplar succeeds fairly well, and imparts an air of grandeur that could hardly otherwise be obtained.

The **Cucumber Tree** (*Magnolia acuminata*).—Few

planters are aware of how valuable this tree is for withstanding the germs and soot of large towns. Experiments have, however, resulted in this highly ornamental tree being added to the list.

The **Tulip Tree** (*Liriodendron tulipifera*).—Excellent examples are not wanting of how valuable a tree this is for towns and streets. It seems to have a wonderful recuperative power, for, scorched, blackened, and encrusted as may appear the falling foliage, yet in the following spring it again puts forth a garb of the freshest and richest greenery. The remarkable four-lobed, truncate leaves render the tree almost without an equal for ornamental planting, while its undoubted smoke-resisting qualities place it high in the rank among town trees.

The **Indian Bean** (*Catalpa bignonioides*).—For various reasons this fast-growing tree is to be recommended for planting in smoky localities. It grows with great vigour in many smoky centres of industry, is a tree of handsome proportions, and, when fully established, flowers freely.

A valuable trait in the character of the Indian bean is that should accident befall it, and the stem get injured, numerous strong suckers are produced, which, as they grow with great rapidity, soon take the place of the original.

The **Common Mulberry** (*Morus nigra*) and the **white-fruited Mulberry** (*M. alba*) may be seen growing satisfactorily in several of the old gardens and nurseries of the metropolis, where they are now buried alive, as might be said, in stones and mortar. That they are excellent town trees will be admitted by every one who sees the fine specimens in Liverpool and Manchester.

The **Honey Locust** (*Gleditschia triacanthos*) is a very suitable subject for planting in smoky localities. In many of the worst smoke-infested parts of London and Manchester are seen goodly specimens of this handsome tree—not poor, miserable trees, but, from their great size, wealth of foliage, and general appearance, betokening perfect health amid their rather adverse surroundings.

The **False Acacia** (*Robinia Pseudo-acacia*)—Almost by the hundred can the false acacia be seen in London and many other English towns, thus proving that it is one of the most valuable trees that we possess for withstanding the injurious effects of an impure atmosphere.

What renders this acacia of greatest value as a town tree is that it retains its rich verdure till well on in autumn. The varieties known as **Decaisneana**, **inermis**, **microphylla**, **macrophylla**, **sophoræfolia**, and the upright-habited are most to be desired.

The **White Beam Tree** (*Pyrus Aria*).—In many of the

THE BEAM TREE IN LONDON

confined spaces in Glasgow the white beam tree grows luxuriously, and produces annually great quantities of its brightly-tinted berries.

The **Lime** (*Tilia europæa*).—Where the situation is not too confined, and where soot and smoke do not abound, the lime may, and does, succeed; but in the worst parts of the metropolis it soon shows signs of distress, the tops of the branches dying off, and the whole tree sooner or later showing the fierce struggle it has to endure with smoke and fumes. As an avenue tree in the more airy and pure parts of a town, the lime has certainly few equals, its general contour and the pleasing shade it affords being points of special recommendation.

The **Sycamore** (*Acer Pseudo-platanus*).—This tree may be classed as amongst the most useful for planting in smoky towns.

In Warrington, where the noxious emanations from alkali and other chemical works are most disastrous in their effects on vegetation, the sycamore is one of the few trees that grow satisfactorily. Being a rapid and strong grower, it is thus seen to be, for a certain time at least, unaffected by its inimical surroundings. The variegated variety would seem from recent experiments to be preferable, and better adapted for smoky localities than the normal form.

The **Weeping Ash** (*Fraxinus excelsior pendula*) would seem to be superior to the common ash for planting in towns. It thrives satisfactorily in many of our largest centres of industry—to wit, London, Liverpool, Glasgow and Manchester.

The **Horse Chestnut** (*Æsculus Hippocastanum*) may be seen in a fairly satisfactory way in many town parks, but only where it is not exposed to smoke and soot to any great extent. In confined spaces both it and the **English Elm** (*Ulmus campestris*) soon show signs of distress, the points of the branches gradually becoming unhealthy, and the trees dying off prematurely. Taking everything into consideration, neither of these trees can be recommended for planting in smoky districts. The **Birch, Walnut, Hornbeam,** and one or two kinds of **Willow** will succeed in the less smoky parts of a town; but they are not to be recommended for planting where the air is constantly impregnated with soot and dust.

The **Mountain Ash,** or **Rowan Tree** (*Pyrus Aucuparia*) and **Bird Cherry** (*Cerasus Padus*) are both good town trees, and excellent examples of both may be seen in the back streets and slums of London.

Sophora japonica is well worthy of recommendation as a tree that is admirably suited for planting in towns. It is of large and rapid growth, with elegant dark green pinnate leaves. Being a native of China and Japan, it may not be perfectly hardy in the northern portions of the British Isles, but it succeeds well in Southern England, and thrives admirably in the most smoke-infested parts of London

Thorns of various kinds succeed well in towns, but they must not be recommended for the most smoky and confined localities.

The **Tansy-leaved Thorn** (*Cratægus tanacetifolia*) is an excellent member of the family for town planting.

Shrubs

Of these there is rather a long list of kinds that are suitable for planting in smoky localities.

Evidently deciduous species possess an advantage over evergreen kinds in the total annual renewal of their leaves, and hence it follows that, as with trees, deciduous shrubs should have the preference.

The following list includes only such kinds as have been proved suitable for town planting :—

Osmanthus ilicifolius is one of the handsomest of evergreen shrubs, and also one of the few that succeed, in a satisfactory way, when subjected to the impurities of a town atmosphere. In the smokiest districts of both London and Liverpool it is unquestionably the best all-round shrub.

Ligustrum coriaceum is a fitting companion to the last, so far, at least, as its powers of withstanding the effects of an impure atmosphere are concerned. Being an evergreen, it is peculiarly well suited for planting in the town garden, where it grows with great freedom.

Olearia Hastii and **O. macrodonta** are excellent shrubs for the London garden.

Aucuba japonica.—This well-known evergreen shrub is of great value for planting in urban districts, it being able to do battle with a more than ordinary amount of atmospheric impurities. For this reason it has been largely planted in town squares and gardens, in the most crowded and densely-populated parts.

Griselinia littoralis.—Although a little-known evergreen, this is well suited for town planting, experiments having proved it to be a most valuable addition to the limited number of shrubs suitable for such a place.

Hibiscus Syriacus is one of the few shrubs that can successfully battle with an impure atmosphere. It is a shrub which town residents should plant freely if they have a bit of ground that they want to look pretty.

The **Warfaring Tree** (*Viburnum lantana*) does not receive that amount of attention to which on its merits it is entitled. It succeeds well in some of the most filthy and smoky districts of our largest cities.

The **Venetian Sumach** (*Rhus cotinus*) is a much-neglected shrub, but for general usefulness it can hardly be surpassed. It is peculiarly suitable for planting in cities.

The **Stag's Horn Sumach** (*Rhus typhina*) must, on no account, be omitted, as it is a shrub of curious appearance and one that thrives well in soot and dirt.

Leycesteria formosa is a capital town plant; this may also be said of the **Flowering Currant** (*Ribes sanguineum*)—indeed, too much praise can hardly be bestowed on these shrubs for planting in the town garden and shrubbery.

Skimmia japonica and the **Snowy Mespilus** (*Amelanchier Botryapium*), too, succeed well in smoke-infested districts; and the various kinds of Lilac—particularly the Common and Persian—have few equals as town shrubs.

In the **Kentucky Coffee-Tree** (*Gymnocladus canadensis*) and **Bladder Senna** (*Colutea arborescens*) will be found two most useful shrubs for the town garden.

Phillyrea Vilmoriniana, Forsythia viridissima, and the **Strawberry Tree** (*Arbutus unedo*) are all more or less suitable for town planting where the atmospheric conditions are not too seriously affected by smoke and dust.

The **Double Furse** (*Ulex europæus florepleno*) is one of our handsomest flowering-shrubs, and a good addition to the list of such as are suitable for planting in town gardens and squares.

The **Spurge Laurel** (*Daphne laureola*) grows freely in many a town garden—indeed, it is no uncommon thing to see large and well-balanced specimens where smoke and filth are the order of the day.

Cotoneasters of various kinds succeed well as town plants. Those to be particularly noted are *C. frigida*, *C. Simonsii*, and *C. vulgaris*.

Euonymus japonicus is another excellent shrub, being almost smoke-defying.

The double-flowered forms of **Prunus sinensis** and the equally ornamental **P. triloba** all succeed well as town plants; while the **Almonds** are quite as good.

Koelreuteria paniculata, the **Laurustinus** (*Viburnum tinus*), **Weigelia rosea, W. amabilis, Deutzia scabra,** the common **Green Box, Gum Cistus** (*C. ladaniferus*), **Mahonia aquifolia, M. Bealii,** and **M. japonica** are all more or less suitable for town gardens, but not for those in the most smoke-infested parts. The **Japan Quince** (*Cydonia japonica*), **Hypericum Nepalense,** and **Euonymus radicans** all do well when subjected to the town atmosphere.

Climbers

Of shrubs suitable for covering walls, trellises, and arbours, and able to resist the dire influences of smoke and soot, there are a few valuable and well-tried kinds.

The **Virginian Creeper** (*Ampelopsis hederacea*) has few equals as a town plant, thriving successfully in the midst of our busiest centres of industry. Many instances could be pointed out in which this handsome climber grows with the greatest freedom in the most impure and smoke-laden atmosphere.

The **Common Ivy** (*Hedera Helix*) is, perhaps, the most valuable of all climbing plants for using in smoke-infested localities. In some of the courts near Ludgate Hill, a district of London that is by no means free from smoke and dust, the ivy climbs houses to a height of 60 ft., and surprises one by its fresh appearance in such localities.

The **Evergreen,** or **Trumpet, Honeysuckle** (*Lonicera sempervirens*) is another shrub of great merit for town planting, as it thrives well in confined spaces, and where the atmosphere is very impure.

Cratægus Pyracantha is a most valuable wall shrub for the town garden. It is of free growth, stands smoke well, and is one of the handsomest berry-bearing plants in cultivation.

Jasminum nudiflorum needs little description, as it is one of our handsomest wall-plants. For smoky districts it is invaluable, blooming freely when flowers are

THE SAVIN, OR JUNIPER, AS A TOWN SHRUB

scarce, and seeming to heed but little the impurities of a town atmosphere.

The **Vine** (*Vitis vinifera*) must not be omitted from our list, it being an excellent plant for withstanding soot, smoke, dust and heat.

Coniferous Trees

Few of these, if any, succeed in a satisfactory way, when constantly subjected to the impurities of a town atmosphere.

Where the conditions are at all favourable the **Austrian**

Pine (*Pinus austriaca*), **Thyiopsis Dolabrata, Toxodium distichum,** and **Cupressus Lawsoniana** do fairly well, but they are not to be recommended for general town planting.

Retinospora plumosa aurea has stood for many years in one of the most smoky districts of Glasgow, and at present looks almost as well as it did when brought from the country, while the **Savin** (*Juniperus Sabina*) may generally be relied upon.

CHAPTER X

TREES SUITABLE FOR HEDGEROW AND FIELD PLANTING

THE well-founded complaint that hedgerow and field trees harbour birds to the serious destruction of the grain crops is largely counterbalanced by their ornamental appearance and the shelter they afford both to man and beast.

In proceeding to consider the trees that are most suitable for hedgerow and field planting four important points must be kept in view. (1) That the spread of branches is, comparatively speaking, small in proportion to the tree's height; (2) that the roots have a downward tendency, or do not ramify to too great an extent; (3) that the tree is well adapted for exposed situations and standing singly; and (4) that the timber value is such as to compensate in some degree for the cost of planting and after-management in the matter of pruning and fencing. Although we rarely find all these qualities concentrated in one tree, still, with careful choice and good after-management, much may be done to produce the desired effect, even in trees of a partially opposite character.

That the wrong class of timber is often planted in our hedgerows and fields is painfully apparent to every one who has paid attention to the subject—trees of wide-spreading habit both in root and branch occupying positions and doing irreparable damage, where others of less obtrusive nature could with advantage have been selected to take their place. Timely and judicious pruning is of the utmost importance in the production of hedgerow timber: but to be productive of the best results, the work must be attended to early, and prosecuted at intervals, as by skilfully shortening the branches from time to time, the spread of root is also checked, thereby securing a double benefit to the farmer and the land.

The **Cornish Elm** (*Ulmus cornubiense*), unfortunately, is far too seldom seen either as a standard or in our woodlands, for which the propagator is greatly to blame, there not being offered anything like a sufficient quantity to meet the demand. As a park or hedgerow tree this distinct and well-marked variety of the elm has much to recommend it—such as a narrow branch spread in proportion to the height, suitability for exposed situations, and the very decided ornamental character it imparts to the landscape when properly placed.

The principal advantages of this tree for hedgerow or field planting are that no pruning is required to keep the branches in bounds, that the spread of branches is very small in proportion to the tree's height, and that the roots do not approach too near the surface, or ramify to any great extent, so as to become injurious to crops in their immediate vicinity.

The **English Elm** (*Ulmus campestris*) is another tree of value for hedgerow planting, and is, perhaps, more commonly used for that purpose than any other. Like the Cornish variety, though in very much less degree, it has an upward inclination, the ramification of its branches being narrow in proportion to its height, while as an ornamental tree and valuable timber-producer it is held in high esteem.

As a shelter tree it is of great value, and as the branches, particularly the lower, seldom spread to a great extent, the injury caused to the adjoining crop is usually not very serious.

The **British Oak** (*Quercus Robur*) is of about equal value with the elm for hedgerow planting—indeed by careful manipulation it is even superior to that tree, being, perhaps, less injurious to the herbage beneath it, and the roots having a greater downward tendency. As an ornamental tree and valuable timber-producer the oak requires no praise from us, these qualities having been recognised from the earliest date.

It bears pruning with impunity, so that all ungainly or far-spreading branches can be cut back or foreshortened at pleasure; while those near the ground, which in any way interfere with the fence or crop, can, by judicious management, be altogether removed. Several of the fastigiate forms of the oak are equally, if not better, suited for hedgerow planting.

The **Lombardy Poplar** (*Populus fastigiata*).—Although of but little value as a timber-producer, yet, as an ornamental tree of singular habit and appearance, the Lombardy poplar is almost unique amongst our hardy deciduous trees.

When planted in judiciously-arranged clumps in the corners of fields, or hedgerows, this poplar produces a most pleasing effect in the landscape: and, being of close, fastigiate growth, it is not in the least injurious to crops in its immediate vicinity. No tree is, however, more readily misplaced than the one in question; and in planting it is well to avoid the prevalent mistake of placing in lines, squares, round or oval forms, or even in single specimens, these methods being highly objectionable and devoid of good taste—that is, if we desire to preserve in the landscape a natural appearance.

The **Sycamore** (*Acer Pseudo-platanus*), as a standard tree, either in field or fence, is worthy of attention, not only for its well-known ornamental character, but also on account of the great value of the wood produced, as it is in this latter respect, perhaps, second to none. As a farmer's tree it is, also, not one of the worst; for, although the branches incline to spread, still, by careful manipulation in the way of pruning, this may be corrected without doing injury to too great an extent to the ornamental qualities of the tree. For imparting both shade and shelter to farm stock the sycamore may be used with the best advantage. Few trees produce such valuable timber when grown singly or in clumps in the corners of fields and paddocks as the sycamore, and no other repays so fully the damage it occasions to fences and the surrounding ground.

The **Hornbeam** (*Carpinus betulus*), for exposed situations and poor soils, has few, if any, equals. The roots do not run near the surface, but, like those of the oak, derive sustenance at a considerable distance from it, and this quality, combined with its somewhat upright inclination of growth and hardy nature, renders it well adapted for hedgerow or field planting, where shelter combined with effect is required.

The **Lime** (*Tilia Europœa*), although one of our most ornamental trees, can hardly be recommended as suitable for situations in which the underlying herbage is at stake. For this latter reason alone, however, can it be omitted from our list, and, as it bears pruning well, does not to any great extent impoverish the adjoining ground. It is at all times a pleasing object in the landscape, and it will therefore be seen that the evil done by shade is in a great degree compensated for.

Amongst coniferous trees, if we except the **Larch** and **Scotch Fir,** few are at all suitable for the end in question. The larch is a much-neglected fence and park tree, this being attributable to an erroneous impression that it is of too stiff and cold an appearance, either for standing singly or giving effect to the landscape. Nothing can, however, be farther from the facts, as when placed so that its fine form is seen to advantage, few deciduous trees are more picturesque than the larch, or offer a better contrast to the ordinary run of our forest trees. It also occasions less damage to the undergrowing herbage than most trees, while, at the same time, it enriches the soil to a great extent by the annual shedding of its leaves.

The **Scotch Fir,** especially for shelter-giving purposes, has much to recommend it for being extensively planted as stock-shelter in the corners of exposed fields.

As it usually rises to a great height without branches, it cannot be considered as extremely injurious to its surroundings, although the shallow-running roots can hardly be spoken of as non-injurious to the greensward.

Planting and Fencing.—Whether for planting in the hedgerow or singly in the fields, good, strong, well-rooted specimens should always be used—indeed, it is well when a home nursery is on the estate to have these specially prepared, by frequent transplantings for a few years previous to their final planting out. The nursery management will require both care and experience, so that trees with strong, fibrous roots equally distributed around the stem may be produced; lanky, ill-grown, and ill-rooted plants having but a poor chance of succeeding under the circumstances. From 10 ft. to 14 ft. will be found the most suitable size for the purpose under consideration. The pits for their reception should be opened of sufficient size to admit the roots without cramping or bending, the bottom and sides being made loose and free with a pick—it will be all the better if the pits have been opened for some time previously to planting, the winter frosts having a beneficial effect in clearing and pulverizing the soil.

This is, however, seldom convenient, as, if in the field, they become filled in, and trampled on by cattle, while gaps in the fences occasioned by these cannot well remain open for any length of time.

The better plan—at least, we have found it so—is to open the pits, plant the trees, and have these fenced in on the same day, as by this method no part of the work has ever to be done a second time, everything being finished up as the work proceeds. In planting, be careful to spread the roots out in an even manner around the stem, as, by so doing, the tree is not only more firmly fixed in the ground, but is enabled to collect food from all quarters.

Fencing should follow up at once the work of planting, as, if the young trees are allowed to remain unprotected for any length of time, they get injured by the farm stock. The fences may be of any desired kind, but, as they are only required for a few years, a simple erection made of small larch poles about 8 ft. long, driven firmly into the ground in a circle around the tree, say, 18 in. from the stem, and made fast to hoops of wood at top, and half-way up, will be found sufficient. Wooden erections are, perhaps, preferable to those made of iron for fencing hedgerow trees, as they seldom require renewing, for by the time the fence has decayed the trees will, in most cases, be out of harm's way.

For shelter and shade clumps in fields probably the best trees to use are the oak and sycamore, the value of timber produced being also a valuable asset.

Pruning and After-Management.—For at least the first ten years after planting, careful and regular pruning of hedgerow and field timber should in all cases be attended to, bearing in mind that timely attention in this way will alone obviate the necessity for heavy prunings at any future stage of the tree's growth.

Early and judicious pruning is necessary to the trees in question ; for it is well known that if branches are allowed to ramify at will, greater injury to the underlying herbage must be committed than where timely pruning and shortening of all straggling branches has been attended to.

Early summer pruning, say, in the month of June, is to be recommended, as at that time, owing to the active circulation of the sap, the wounds heal up much more quickly than when the operation is performed at any other season of the year. The pruning should be performed by a person who is thoroughly conversant with the work, haphazard cutting and hewing, by an inexperienced hand, and at any season, being injurious. If the young trees have been well attended to in the matter of pruning whilst in the nursery border, little or no attention will afterwards be required—at least for a number of years.

The main object in pruning both hedgerow and field trees is to develop a valuable main stem which is only to diverge into branches at a given height from the ground, and to prevent the overgrowth of straggling branches farther up, so as to maintain a symmetrical and rather fastigiate head.

CHAPTER XI

ORNAMENTAL PLANTING

As the beauty and picturesqueness of an estate are so dependent on the disposition of its single trees, groups and plantations, the forester should never lose sight of the fact, even when dealing with plantations that are mainly intended for the value of the timber produced.

In ornamental planting one of the principal things to bear in mind is to allow sufficient space for each of the permanent standards to develop its true and natural character. Instead, therefore, of planting indiscriminately and in a sort of haphazard way, have a fixed idea, and only plant on a regular and well-matured plan. Should the individual standards be considered stiff or unsightly for the first few years, a good plan is to fill up the spaces between them with small-growing trees and shrubs, these being removed subsequently as necessity demands, but always before damage to the permanent specimens has been brought about by too close contact. The habit of the tree or shrub, and the size to which it will ultimately attain are points that should never be lost sight of in ornamental planting. Far too often the mistake is made of planting specimen trees too near roads and buildings, or, quite as bad, too close to each other. This is a most unfortunate mistake, as it sooner or later necessitates the sacrifice of specimens when their full beauty is probably developed, or when they can least be spared.

In the case of both hardwooded and coniferous, the latter in particular, the trees are planted when young and small, and ample room is supposed to be given to them ; but, as they grow up and near perfect development, the spread of branches soon points out that a great mistake has been made in the space of ground allotted, and that the ultimate size to which the particular specimen would attain had never, at the time of planting, been duly considered. There is little left then to be done but either to put up with crowded and ill-formed specimens, or to attempt their removal to more suitable quarters. The latter, in the case of trees that have been left undisturbed for many years, and when tap and side roots have been sent down to a great depth for safe anchorage, is quite out of the question, the only remaining way out of the evil being to sacrifice the specimen by having it stubbed out and removed. In any case, in the event of one tree encroaching too closely on another, no time should be lost in deciding which is to be removed, for, if they crowd into and interfere with each other, the branches on that side will get thin and die off, and the tree will thus lose the uniformity of appearance on which its whole beauty depends. Should crowding of specimen trees be permitted for even a few years, partial disfiguration of the trees will have been brought about, and it will take years of careful management to restore them to their original beauty.

In order to plant with any degree of certainty as to future results, it is absolutely necessary to become first of all well acquainted with the nature and habit of every tree planted, as also the properties and peculiarities of the various soils, and aspect and position of the ground to be planted.

Probably no other branch of forestry requires sound judgment and correct ideas to such an extent as the laying-out of parks and grounds, or, in other words, landscape and ornamental planting—and this knowledge can only be attained by perseverance, investigation, and study.

In selecting sites for the various trees, it will be well to bear in mind that certain species are better suited than others for planting in exposed places, for using in particular soils, and for inserting in either dry or damp ground.

By the lake or pond side the **Deciduous Cypress** (*Taxodium distichum*), the **Bhoton Pine** (*Pinus excelsa*), the beautiful cut-leaved **Imperial Alder** (*Alnus glutinosa imperialis*), the **Golden** and **Purple Willows** may all be planted with the best chances of success ; whereas none of these would thrive well on dry or sandy soils. Should chalk or calcareous soil crop up, we have good subjects in the **Fern-leaved Beech** (*Fagus sylvatica asplenifolia*), the **Cephalonian** and **Spanish Silver Firs** (*Abies cephalonica* and *A. Pinsapo*) and many of the **Pyrus** family ; while, where only a small quantity of loam overlies the gravel, the **Scotch** and **Cluster Pines** (*Pinus sylvestris* and *P. Pinaster*), the pretty and far from common **Manna Ash** (*Fraxinus Ornus*), and many others may be successfully planted. In good, rich soil, and where the position is fairly sheltered, there are no end of ornamental trees, both hardwooded and coniferous, that may be planted. Some of the **Magnolias,** but particularly *Magnolia stellata*, *M. acuminanta*, and *M. Umbrella* should find a place, while the **Juneberry** (*Amelanchier canadensis*), the **Cornelian Cherry** (*Cornus Mas*), and various **Thorns** should not be neglected. In exposed places plant clumps of the **Corsican** and **Austrian Pines** (*Pinus laricio* and *P. austriaca*), following up with the hardy **Spruces** and **Cypresses.** Such dainty conifers as the **Japanese Cryptomeria** (*C. japonica*), the **Elegant Cryptomeria** (*C. elegans*), **Fitzroya patagonica, Prince Albert's Fir** (*Tsuga Mertensiana,*) and **Umbrella Pine** (*Sciadopitys verticillata*) must have cosy corners and good soil to show their beauty to perfection.

Pitting and Planting.—The pits for the reception of ornamental trees should be well formed : that is to say, should be made of sufficiently large size for the specimens to be planted—indeed, it is always preferable to dig out pits of a greater size than are required, thus allowing of a quantity of broken-up soil being placed beneath and around the roots. In any case, have the bottom and side of each pit thoroughly broken up, and should the soil be found to be of inferior quality, it is best to substitute that of a more desirable kind such as is known to be suitable for the wants of the particular specimen. In removing large specimen trees great care should be exercised—first, that the roots are uninjured, or if at all, in a very small degree ; second, that a good ball of earth is attached; and, third, that replanting is delayed as short a time as possible. Have the pit, for the reception of the tree, dug out and prepared before the specimen is lifted. Lift carefully by undermining the roots, and digging out a deep trench at a reasonable distance from the stems, the distance from the stem to be regulated by the root-spread, size of the specimen, etc. In order to avoid falling apart, the ball of earth should be bound with a strong mat or tarpaulin, the ends being made fast around the stem.

A stout low-wheeled truck has been found a most convenient appliance for removing the specimen; but in the case of large and heavy trees the common timber wheels, or janker, may be found of great service. The truck to which we refer should be specially made, and should be of the following dimensions: length, 5 ft.; width, 3 ft.; on wheels 1 ft. 3 in. in diameter, these being placed so as to be below the level of the body, thus avoiding contact with the load when that is either longer or broader than the specified dimensions. The framework should be of stout oak, and the bottom 3 in. thick boarding. The timber wheels, or janker, being an adjunct of forest appliances, is always at hand, and does away with the necessity of procuring any of the elaborate and costly carriages usually recommended for transplanting large trees and shrubs. The tree being placed in the pit opened for its reception, the roots should be spread out and the soil replaced and trampled firmly, and afterwards well watered.

Staking the Trees.—Next to careful planting and watering a matter of the most vital importance—but one that, unfortunately, is too often lost sight of—is the efficient staking or otherwise securing of large transplanted trees. The great strain and consequent damage to the roots of large transplants when allowed to rock about with every gust of wind is not only highly injurious, but, viewing the matter from a point of neatness, few things in forestry have a more unsightly or neglected appearance than trees almost blown over by the wind. Various are the methods usually adopted in staking and tying newly-transplanted trees.

As no hard and fast lines can be laid down, as a rule, the size of the trees and the exposure of the positions in which they are planted must determine the method of procedure to be adopted. It may be that when growing in a low-lying, sheltered valley, trees of even 8 or 10 ft. in height may be perfectly safe without stake or tie of any kind, whereas others of similar or even smaller size, growing in an open position and exposed to the prevailing winds of the district, will require a mooring of the most secure description, and adjusted in the most efficient manner. For trees and shrubs up to 6 ft. in height, a strong, sharp-pointed stake should be driven firmly into the ground, within about 9 in. of the main stem, and on the most exposed side. The stake should be fully 7 ft. long, and driven, not perpendicularly, but with the head slightly inclined from the tree, and in the direction from which the prevailing winds usually blow. A strong ligature of some kind, such as tarred rope or thick matting, is then placed round the stem of the tree requiring support, at about 4 ft. from the ground, and made fast to the stake at a similar height.

By crossing the tie between the tree and stake, a more efficient job results, as there is then less room for the tree to work in when rocking with the wind. It will thus be seen that the nearer to the stem of the tree the stake is placed, the greater will be the power of resistance.

When the trees and shrubs are from 8 ft. in height and upwards, but especially in the case of evergreen species, stronger moorings than those just described will have to be employed, and the following plan we have found suitable for almost any emergency. A strong band of leather, or several strands of tarred rope, are placed loosely around the stem of the tree to be staked, and at, say, three-quarters of its height. Three wires—ordinary fencing wire does well—are then joined to this collar, two on the most exposed side and one on the other, and made fast to stout stakes driven firmly into the ground, at a distance from the main stem proportionate to the tree's height. In the case of very large trees, or those that have been reinstated from a fallen position, double wires are used in a manner similar to that just described, the collar, however, being unusually strong. The advantages of double wires are extra strength, and the fact that they admit of being twisted at any time to the tightness required.

The above may be considered the two principal ways of staking large transplants, but occasionally cases will crop up in which it may be necessary to resort to other methods, but such are peculiar cases, and must be dealt with in a peculiar manner.

Cut-Leaved, Weeping and Fastigiate Trees

These have their own place in landscape gardening, and when placed in the hands of a skilful planter are capable of producing the most beautiful and pleasing effects. The majority are graceful in outline, distinct and impressive in appearance; in fact, possess all those qualities which render them specially valuable for the embellishment of park or garden. Great care is, however, necessary in planting these trees, for if not used with the greatest discretion, the good effects they are so capable of producing in the hands of a trained planter are destroyed.

Weeping Trees.—Both picturesque and beautiful is a well-developed specimen of the Weeping Beech, and particularly so when associated with trees of a light and airy appearance, for the general character of the tree is somewhat massive and lumpy. Being of large growth, the Weeping Beech wants plenty of room for development. There is also a desirable weeping form of the purple Beech.

The Kilmarnock Weeping Willow (*Salix caprea pendula*) is one of the most popular and widely cultivated of weeping trees. It originated near Ayr, in Scotland, and was so named to distinguish it from two other well-known varieties, the common Weeping Willow and the American Weeping Willow—two desirable forms for waterside planting. In the American Fountain Willow we have another excellent weeping tree, while the ringed-leaved Willow (*S. Babylonica annularis* or *crispa*) is one of the most curious and picturesque of the whole group.

Amongst the Birches are at least two desirable weeping forms in the cut-leaved (*Betula pendula Youngi*), Young's weeping Birch, and the weeping white Birch (*B. alba pendula*), both of which are admirably adapted for lawns, even where space is restricted. The Birch will thrive on poor, light soil, and is, therefore, peculiarly suitable for planting in gravelly or rocky formations.

The distinct habit of the Weeping Mountain Ash, or Rowan tree, has made it a favourite in the embellishment of small grounds, and the wealth of conspicuous fruit produces a telling effect, especially if the tree is backed up by darker foliaged trees or shrubs. There are other forms of Pyrus well worthy of attention, such as the pendulous Siberian Crab (*P. prunifolia pendula*) and *Salicifolia pendula*, the latter in particular being a distinct and graceful small growing tree. The Weeping Mulberry (*Morus alba pendula*) is occasionally seen in good form when it is a desirable acquisition, so is the Weeping Laburnum (*L. vulgare pendulum*) and the curious and scarce *Gleditschia triacanthos excelsa pendula*. In the silver-leaved Weeping Holly we have one of the best of evergreens for confined situations; and *Cornus florida pendula* is a strikingly beautiful tree in which the weeping habit is decidedly pronounced.

Among Weeping Elms some varieties are most pronounced, but there are several others all equally worthy of culture. Of the English Elm, probably the best is *Ulmus campestris Petersii pendula*, while of the Mountain or Scotch, that known as *U. montana pendula* is the most desirable. Weeping thorns are by no means common, but they are probably the most natural and graceful of all drooping trees of modern growth. In the flower garden at Regent's Park is growing a beautiful specimen of this tree.

Perhaps the most common of weeping trees is the Weeping Ash (*Fraxinus excelsior pendula*), and its strong, vigorous growth causes it to be employed where others would not succeed. Even in smoky localities it is quite at home, as many fine old trees throughout London clearly prove that the deleterious effects of an impure atmosphere have but little effect on its hardy constitution. It is one of the best of weeping trees for forming an arbour, and as a distinctly ornamental specimen for the park or large lawn it has few equals.

Amongst coniferous trees we have several distinct and beautiful pendulous trees or shrubs, such as *Abies excelsa inverta* and the well-known and much appreciated *A. Smithian* or *morinda.* The pendulous Hemlock spruce (*A. Canadensis pendula*) is one of the prettiest and most natural in appearance of all weeping trees.

There are several distinct and beautiful forms of the Cypress, particularly *C. Lawsoniana gracilis pendula* and *C. Lawsoniana alba pendula.* The Weeping Chinese Juniper (*Juniperus Chinensis pendula*) and *Virginiana pendula* are well worthy of attention where this particular class of trees or shrubs is in request. Than the weeping form of the common Larch, perhaps no tree is more beautiful, the feathery peagreen foliage being distinct from that of every other tree.

Cut-leaved Trees.—Of the Maples we have several cut-leaved forms, such as the beautiful *Acer palmatum laciniatum* and *dissectum,* while of the Norway Maple, the variety known as *Platanoides dissectum* is particularly valuable on account of the freely divided and beautifully coloured leaves.

In the cut-leaved Birch (*Betula alba laciniatum pendula*), we have at once one of the most graceful and distinct of hardy trees. Not only are the leaves cut up and divided, but the weeping sprays of foliage, owing to the fine, whip-cord-like branchlets, produce a most beautiful and graceful effect. In addition, this Birch is a tree of neat, clean growth, and will succeed in a satisfactory way even on poor gravelly soils and upland situations.

Amongst the Beam trees (*Pyrus*) we find several good cut-leaved forms, such as *pinnatifida* and *salicifolia,* both highly desirable medium-sized trees and valuable for woodland margins or steep, undulating grounds. Numerous forms of the Oak have appeared with incised leaves, such as *Quercus dentata* and the fern-leaved variety of the Turkey Oak (*Quercus Cerris asplenifolia*), a decidedly beautiful and worthy form. Of the English Oak (*Q. Robur*) we have at least two good cut-leaved forms in *Pedunculata asplenifolia* and *pectinata.* There are other varieties of the Oak in which the foliage differs considerably from that of the species in being much more freely divided.

Rhus glabra laciniata has become widely known of late years on account of its neatly divided foliage and as being a good town shrub. The leaves are of a delicate pea-green and the flowers inconspicuous and succeeded by feathery lawns.

The common Alder has at least two forms that are much sought after for dampish ground by the pond or lake side, in *Alnus glutinosa laciniata* and its well-marked and distinct variety *imperialis.* Both attain to a goodly size, and when suitably placed where the abundant, rich green, deeply divided leaves can be readily seen, are distinctly valuable for contrast and waterside effect.

The incised or cut-leaved Hornbeam (*Carpinus Betulus incisa*) is rarely seen, but it is of so distinct a type that its inclusion in this list is considered desirable. In this case the almost entire beech-like leaf is finely divided and the whole tree presents a comparatively light and airy appearance. Of the common Hawthorn there is a specially elegant cut-leaved form in that named *laciniata*, but its distinctive characteristics are best revealed when planted in conjunction with the species.

Several of the Vine family are rendered highly ornamental for covering walls and pergolas by reason of their beautifully coloured and divided leaves. *Vitis heterophylla dissecta* is one of the best in this way, but others of the less-incised forms are well worthy of attention. *V. vinifera purpurea* has much to recommend it as an ornamental-leaved variety, and is specially suitable for covering mounds of earth, stones or tree stumps.

We must not omit to include the cut or fern-leaved variety of the Lime (*Tilia platyphillos laciniata*), which for neat habit of growth and distinct character from the species merits special attention. The cut-leaved Japan Maples are exceedingly beautiful, but their slow growth and difficult propagation will always be against extensive planting, and render them rare and expensive. Other interesting cut-leaved trees are the willow-leaved Ash, laurel-leaved Willow, scarlet Oak, Maidenhair tree, and that most beautiful of all, the fern-leaved form of the golden Elder.

Though somewhat stiff of outline the cut-leaved Horse Chestnut (*Aesculus*), in the form known as *laciniatum*, is not to be despised, and affords a striking contrast to the species.

The fern-leaved Beech (*Fagus sylvatica asplenifolia*) is probably the most commonly distributed of cut-leaved trees, and rightly so, for it is certainly one of the most beautiful and distinct of the type. *F. sylvatica quercifolia*, the Oak-leaved, is also worthy of attention, but cannot compare with the former in point of beauty. Even of the common Walnut (*Juglans Regia*) there is a cut-leaved variety, which is distinct and worthy of culture as an ornamental tree. It is named *laciniata* and is fairly common.

Of the numerous varieties of the Sweet or Spanish Chestnut, the most desirable is that which bears the rather cumbrous name of *Castanea vesca heterophylla dissecta*. It is one of the most beautiful of hardy trees, in which the long, narrow leaves of the upper parts of the branches droop in a graceful manner and render it so much sought after in ornamental gardening.

Fastigiate Trees.—The Lombardy Poplar (*Populus fastigiata*) is at once one of the most conspicuous and picturesque of tapering trees. Though it can hardly be described as a beautiful tree, yet for landscape effect the tall, spiry, column-like appearance renders the Lombardy Poplar of particular value for certain well-chosen positions in our parks and grounds. In the neighbourhood of a town or country village it produces an effect almost akin to architectural embellishment, while in flat or low-lying districts, and particularly near water, it is most at home and probably looks best. It also associates kindly with old ruins and has a pleasing effect when rising out of pointed-headed Cypresses or Yews, but in all cases it is a tree that should be used sparingly and with extreme caution. Forming avenues or lines of the tree is not good taste, while planting single specimens in open situations should be avoided.

Of the Poplars there are several upright-growing varieties in addition to the Lombardy, such as the beautiful *P. alba Bolleana*, a desirable fast-growing tree with a character of its own.

The Oak, too, has its decidedly upright form in *Quercus pedunculata fastigiata*, which in old parks is a fairly common tree. A decidedly beautiful small growing tree is the upright form of the dwarf Acacia (*Robinia Pseudo-acacia inermis fastigiata*) which for confined spaces and wealth of delightful pea-green foliage is a valuable small-growing variety.

The Elms are not wanting in upright-growing forms, both the English and Scotch being represented, the former in *Dampieri aurea*, and the latter in *Montana fastigiata*. A beautiful and distinct tree of upright habit will be found in the fastigiate variety of the Tulip tree (*Liriodendron tulipifera fastigiata*).

Cupressus sempervirens, in many parts of England, but particularly by the coast, forms a beautiful tapering evergreen tree with the darkest-green foliage. For clump planting it is particularly desirable, but it is not quite hardy in some parts of the country. Another coniferous tree of naturally erect growth is the red or Virginian Cedar (*Juniperus Virginiana*), which in sheltered sites is a most desirable member of the family. Two other Junipers are of decidedly strict growth, *J. drupacea* and *J. thurifera*.

The Irish Juniper (*Juniperus hibernica*) originated in Ireland, most probably as a chance seedling. It is a most desirable shrub, and on account of its singular habit of growing in a compact, slender and graceful column, has proved itself a most valuable evergreen for almost any position, but particularly where geometrical gardening is carried out. The silvery glaucous hue of the thickly produced foliage is extremely beautiful. The nearly allied but much smaller growing *J. hibernica compressa* forms a compact slender pyramid of bright-tinted foliage and is an excellent plant for rockwork or small gardens.

Amongst the Cypresses are several upright-growing forms, probably the best, and certainly the most widely distributed, being Lawson's erect Cypress (*Cupressus Lawsoniana erecta viridis*), of dense, erect habit and with vivid green foliage. *Cupressus macrocorpa lutea*, raised by Messrs. Dicksons, of Chester, has a decidedly neat and upright habit of growth, while the distinct golden tint of the foliage renders it a desirable acquisition for ornamental planting.

Than the Irish Yew (*Taxus baccata fastigiata*) with its dark sombre foliage and neat columnar outline, few evergreens are more conspicuous and effective. This favourite and far-distributed variety originated at Florence Court, in Ireland, a century and a quarter ago.

Cupressus macrocarpa fastigiata is a well-marked variety in which the branches closely press to the main stem. The foliage is of the brightest green, and the long whip-cord-like shoots with the conspicuous reddish back, impart a peculiar grace to healthy specimens. *C. torulosa*, for planting where space is confined, is a decided acquisition, the easy though columnar habit of growth, slender branchlets and bright glaucous foliage being all points of recommendation.

The upright form of the Birch (*Betula alba fastigiata*) has a peculiarity of habit that, associated with the slender weeping shoots, is highly ornamental; while *Cornus stricta* and the upright variety of our common Hawthorn have both points of recommendation.

CHAPTER XII

TREES BEST ADAPTED FOR VARIOUS SOILS

THERE is, perhaps, no soil so bad and barren that it may not be rendered either profitable or ornamental by judicious planting; but, as might be expected, there is often a great want of knowledge as to the proper kind of trees to be chosen to suit a particular soil. In looking over a large extent of woodland one will often be struck with the great disproportion in size of the individual trees of a species; but it will generally be noticed that where the largest and healthiest occur the tree is usually growing upon its own soil, and is found to be flourishing at the expense of all around it. Thus the finest oaks will be found where the soil is deep and loamy, resting on clay; beech, and the Austrian pine (*Pinus austriaca*) upon a calcareous gravel, resting on a bed of chalk; ash and elm on a dampish, loamy gravel; birch in a light, black loam, with a gravelly substratum; Spanish chestnut, in a good loamy or gravelly soil, not too damp; the Scotch and Corsican pines (*P. sylvestris* and *P. laricio*) at fairly high altitudes, and in gravelly, well-drained soils; and the Cluster and Aleppo pines (*P. Pinaster* and *P. halepensis*), in almost pure sand on the sea-coast. Some trees grow rapidly for a few years in almost any soil, but where the soil is unsuitable they generally show signs of distress after a time, make little or no progress, and ultimately become stunted and ill-grown. Should the soil be very unfavourable, they die outright. Instances of this are, unfortunately, far too common wherever one travels over the country, owing to trees of a kind that are utterly unfitted for the particular class of soil being planted in a haphazard kind of way, without any consideration of their individual requirements. For all practical purposes with reference to tree-culture, soils, generally speaking, may be divided into six distinct classes—peaty, chalky or limey, gravelly, clayey, loamy, and such as contain ironstone, coal, etc.

(1) **Peat.**—Few trees will succeed well on an unreclaimed peat bog, but, where draining and soiling have been attended to at the outset, the number that grow and produce a fair amount of valuable timber is almost without limit among our generally cultivated species. Among conifers that have proved themselves suitable for bog planting are the larch, Scotch pine, and common and black spruces (*Picea excelsa* and *P. nigra*). The larch grows rapidly, and is less subject to disease on peaty than any other soil—indeed, up to a few years ago, I cannot remember having seen a trace of any of the diseases which have rendered the life of the tree so precarious of late years in this country. In thinning a larch plantation of fully sixty years' growth I found the trees felled to be perfectly healthy, and of exceptional quality, with, on an average, 72 ft. of wood in each. The subsoil in this case was clay, and the bog, previous to being planted, had been cut over for fuel. The Scotch pine grows almost as freely as the larch—the average in over fifty trees measured being about an eighth less—under similar conditions. Natural reproduction of the Scotch pine goes on so rapidly that it must be considered one of the very best trees for planting on peat bog. The spruces are excellent trees for planting on reclaimed peat bog, where they produce a fair amount of timber and afford excellent shelter to other trees. Of hardwoods, the beech is one of the best for bog planting, as it grows rapidly and produces a large amount of clean timber. The alder grows luxuriously on peaty soils, and shows no traces of disease or canker. Another excellent bog tree is the Gean or Wild Cherry, and this may likewise be said of the holly. Ash and oak are not generally of large size, nor are they always healthy on peat bog, even when it has received a great amount of attention in the way of reclaiming. Birch, lime and poplar

of various kinds are all suited for planting on well-drained bog.

Among coniferous trees, a large number are well suited for planting in reclaimed peat bog. By way of experiment I have planted specimens of various kinds in newly-formed plantations, and in nearly every case the trees have grown well, particularly when partially sheltered. *Cupressus macrocarpa* is one of the best, and not one whit behind it are *C. Lawsoniana* and *C. Goveniana*. *Wellingtonia gigantea* and *Sequoia sempervirens* have done well, while *Pinus laricio* and *P. austriaca* grow freely. I find that the majority of the recently introduced conifers do well on prepared peat bog—that is, where a quantity of loam has been incorporated with the bog and all superfluous moisture drained away.

(2) **Chalky Soils.**—The beech is peculiarly well suited for planting in chalk districts, for it will grow and produce a large quantity of excellent timber where but a few inches of loam overlie the chalk. It is a fact that, in Southern England particularly, in order to find where the chalk beds lie, one has only to be guided by the line traced out by the largest and most luxuriant beeches.

The beech will grow freely enough on almost pure chalk, but it certainly flourishes best where loam, say, from 1 ft. to 3 ft. in depth, overlies the chalk, or is incorporated with it, as on the Chiltern Hills.

The Norway maple (*Acer platanoides*) revels in a chalky soil, and so does *A. colchicum rubrum*. These are both handsome, hardy, large-growing trees, and well suited for extensive forest-planting under certain conditions of soil. White poplar (*Populus alba*) is an excellent tree for planting in chalky districts—indeed, it is surprising to see to what an immense size it attains on almost pure chalk.

Other poplars that do almost equally well on the chalk formation are *P. monilifera* and *P. canadensis*, both excellent, free-growing trees. Elms, particularly the Huntingdon and the American, grow rapidly, and attain to a large size, where but a small quantity of loam is present in the chalk. The wych elm grows freely in chalky districts, and this may also be said of the common and silver-leaved lime. False acacia (*Robinia Pseud-acacia*) is an excellent tree for chalky soils, and there attains to a greater size than in even the richest of loams. The alder and birch also thrive with vigour on chalky soils. Indeed, most trees which in a state of nature grow in damp or marshy soils, are well suited for planting where chalk is the component of the main soil, and this is explained as follows :—Chalk, although sufficiently porous to allow water to percolate through it, has, like all other calcareous matter, a strong attraction for water, and acts like a sponge in holding it in considerable quantity for a very long time. Among the conifers that are suitable for chalky soils the Spanish fir (*Abies Pinsapo*) is one of the best. In the chalky districts of Southern England it thrives with unusual luxuriance. The Mount Enos fir (*Abies cephalonica*) is, likewise, well adapted for growing in chalky districts. Of evergreen trees that succeed well on chalk the number is well known to be limited, and it is important that two such beautiful conifers as the Spanish and Mount Enos firs should there find their most congenial home.

Both the Scotch and Weymouth pines (*Pinus sylvestris* and *P. Strobus*) are well suited for planting on chalk, and many fine examples of both may be seen on the chalky reefs of Kent and Surrey.

The common Yew grows freely where hardly a particle of soil overlies the chalk formation.

The Giant Arborvitæ (*Thuja gigantea*) is peculiarly suitable for planting in chalky soils : and the Lebanon Cedar (*Cedrus Libani*) is never found in greater perfection than when growing in the chalk with a fair depth of loam atop. *Wellingtonia gigantea* also does well.

(3) **Gravelly and Sandy Soils.**—The Corsican pine is an excellent tree for planting on gravelly soils, and some of the largest and finest specimens in this country are growing in a disused gravel-pit, and this may also be said of the Douglas fir (*Pseudotsuga Douglasii*).

The Scotch pine is well known to be one of the best conifers for planting in gravelly soils, where it reproduces itself in great numbers, when the conditions of growth are at all favourable.

Pinus Pinaster, the cluster pine, is, perhaps, one of the most valuable conifers for planting either in gravelly or sandy soils. The great value of the tree in reclaiming sandy tracts, both at home and abroad, has been so often described that further reference here is not required. The Aleppo pine is a good companion to the Pinaster, and grows with great freedom in a sandy or gravelly soil, within the influence of the sea. Gravelly soil also suits the Weymouth pine, on which it produces a fair quantity of very resinous timber. Both the beech and oak produce a large volume of timber on poor gravelly and sandy soils.

(4) **Clay Soils.**—The soil here referred to is genuine clay, devoid of stones, and without a particle of sand or loam in it.

It occurred on the slopes, and for some considerable distance along the sides of one of the park roads on an estate in England.

This is recorded simply to show what species of trees are best able to succeed when planted in pure clay. The pits, in this case, it may be well to mention, were dug and the soil thrown loosely up for a month previous to planting, but no soil was added to the stiff clay. Nearly one hundred kinds of trees and shrubs were used, but out of all these not more than eight are doing well, the others having gradually died out, or become so rusty and miserable looking that their removal was compulsory. First among the trees that have succeeded is the giant arborvitæ (*Thuja gigantea*), which seems to revel in what is generally considered the most unkindly of soils. *Cryptomeria japonica* has also done well, but the trees of this kind, though bushy and well furnished, have grown at a comparatively slow rate.

Cupressus macrocarpa also has done fairly well: the growth certainly has not been rapid, but for all that the general appearance of the trees is the reverse of what one might expect from the unfavourable nature of the soil. The Indian Cedar (*Cedrus Deodara*) we have found to be peculiarly well suited for planting in clayey soils, the bright silvery tint that is so characteristic of this cedar when well grown being discernible in the clay-grown specimens. *Pinus austriaca* has, in a few instances, done well, the foliage being ample and of the usual dark yew-green. Amongst shrubs the double-flowered gorse (*Ulex Europœus*, fl. pl.) has done best of any—indeed, it has grown and increased freely, and would seem to be quite as much at home as in its natural element—a dry, gravelly bank. These may be considered as the trees that have succeeded best in stiff, clayey soil. Few of the pine tribe did well, and this may also be said of the spruces, cypresses, yews, junipers, arbutus, dogwood, cotoneaster, hollies, and others planted.

It is, of course, far from advisable to plant trees or shrubs in such unkindly soil without first adding other of better quality; but it is of great value to know that there are a few trees and shrubs that will thrive almost in defiance of the stubborn and unkindly nature of a stiff clay soil.

(5) **Ironstone Soils.**—The particular class of soil to which I refer, and which in several districts occurs in plenty, is on the coal and ironstone formation, where the top soil is usually shallow, and the subsoil consists of a loose, yellowish rag that is largely impregnated with iron. In most places but a very small quantity of soil exists, and that is of the poorest description, varying in depth according to the lie of the measure. The Spanish Chestnut is one of the very best trees for this soil, growing with freedom, and producing a fair amount of good timber, while its appearance indicates perfect health.

Birch and beech do well, although neither of them attains to a large size. The latter reproduces itself freely from seeds, and soon spreads wherever a footing can be got. Sycamore grows freely, particularly where the pan is broken up, and produces a small quantity of good timber. Another tree that seems perfectly at home on the coal and ironstone is the wild cherry, for there it grows to a fair size, flowers freely, and produces excellent timber. Larch cannot be recommended for this soil, but in places where a small quantity of loam overlies the coal and ironstone it grows with great freedom for a number of years, and the timber, if cut early, is of good quality. The common spruce soon dies out, although it may grow freely enough for a number of years after being planted, and wear a healthy appearance. Oak and ash do fairly well, but they rarely attain to a large size or produce first-class timber. Rhododendrons almost revel in this soil, and some of the largest and healthiest are growing with their roots in close contact with the coal and ironstone.

Shrubs for Hot and Dry Soils.—The Bladder Senna (*Colutea arborescens*) is one of the most useful of shrubs for planting in poor, hot, dry soils, and not only will it succeed and flower well in these, but it is equally valuable for using where the air is chemically impure, and for that reason has few equals for the town or city shrubbery. *C. cruenta* is also valuable in a similar way. The Sea Purslane (*Atriplex halimus*) is another valuable shrub for planting in hot and dry situations as is *Caragana arborescens*, the native Barberry (*Berberis vulgaris*), many forms of Genista and Cytisus, *Spartium junceum*, and the double flowering Gorse. Another excellent shrub is the Box Thorn or Tea Tree (*Lycium Europæum*) which is useful for covering an arid hot bank, several species of Cotoneaster, particularly *C. horizontalis* and *C. microphylla*, and the neat and curious *Muchlenbeckia complexa*. Helianthemums also do well, so does the Venetian Sumach (*Rhus cotinus*), and several varieties of bramble, but especially the double pink flowering form. The Rest Harrow (*Ononis arvensis*), a native shrubby plant of great floral beauty, also does well; while the little known *Celastris articulatus* should not be neglected in hot and dry situations. Other good shrubs for hot dry banks are *Potentilla fruticosa*, the Tamarisk and *Juniperus tamariscifolia*.

Concluding Remarks.—In conclusion, it may be pointed out that it is only by a careful selection of soil that we may expect tree planting to be successful, and I have no hesitation in saying that many failures can be clearly traced to errors of judgment in the selection of trees for planting on particular soils. The subject is a wide and complicated one, and it must be admitted that very perplexing diversities occur with the same kinds of trees on what, to all appearance, is the same class of soil. There are, of course, other considerations beyond the soil itself which must be taken into account, such as aspect, elevation, and whether the ground is inland or on the coast.

With reference to some of the newer conifers it must be admitted that soil and situation have a wonderful influence on their successful culture, and this applies in particular to such kinds as are not perfectly hardy and liable to injury by unseasonable frost. The too common practice of selecting warm and sheltered spots for such is, in the main, to be condemned.

It may be said that *Abies cephalonica* and *A. Pinsapo* are not worth growing, and in many places they are not, but when planted on limestone or chalk they are highly ornamental.

The same holds good in the case of *Tsuga Mertensiana*, *Picea sitchensis*, and *P. excelsa*, which are rarely seen in good form except on soil of a peaty description.

Every one at all interested in trees and shrubs knows that there are certain kinds which in a state of Nature are only found growing in a peaty soil, mixed it may be more or less with sand, and any attempt to cultivate them in other soils is productive of very unsatisfactory results.

Who would ever think of planting the so-called American or peat plants, Cape Heaths, etc., amongst gravel or chalk, or Rhododendrons where lime is present in the soil? And these facts show us that there is something in the composition of certain soils only suitable for the requirements of a certain class of plants.

Another curious fact is this, that when growing on certain soils the timber of one species of tree is found to be far more durable than in others.

Deep loamy soil and soft peat produce timber that is usually of a second-rate description, being deficient in firmness.

By studying the geological strata of a district much useful information may be learnt regarding the trees best suited for planting.

The following alphabetical table will serve to show at a glance the trees that have been found best suited for planting in the class of soil under which they are enumerated :—

Reclaimed Peat Bog.

Hardwoods.

Alnus glutinosa
——— ——— imperialis
——— ——— laciniata
Betula alba
Cerasus Padus
——— vulgaris
Fagus sylvatica
——— ——— purpurea

Populus alba
——— balsamifera
——— canadensis
Quercus Robur, and vars.
Salix fragilis
Tilia europeæ
Ulmus alata
———montana

Conifers.

Abies concolor
——— bracteata
——— nobilis
——— Nordmanniana
Cedrus Deodara
Cryptomeria japonica
Cupressus Goveniana
——— Lawsoniana
——— macrocarpa
Juniperus chinensis
——— recurva
——— Sabina
Larix europeæ

Larix Kæmpferi
Pinus austriaca
——— excelsa
——— laricio
——— sylvestris
Retinospora ericoides
——— plumosa
——— ——— aurea
Taxus baccata
Thuja gigantea
——— occidentalis
Thujopsis borealis
Wellingtonia gigantea

Chalky or Calcareous.

Hardwoods.

Acer colchicum rubrum
——— dasycarpum
——— Negundo
——— platanoides
——— Pseudo-platanus
Æsculus Hippocastanum
——— rubicunda
Alnus glutinosa, and vars.
Amelanchier Botryapium
Amygdalus communis
Betula alba
Castanea vesca
Catalpa bignonioides

Cerasus Padus
Cratægus (nearly all)
Cytisus Laburnum
Fagus sylvatica
——— ——— purpurea
Fraxinus excelsior
——— Ornus
Gleditschia sinensis
——— triacanthos
Koelreuteria paniculata
Populus alba
——— balsamifera
——— canadensis

CHALKY OR CALCAREOUS—*continued.*

Hardwoods—continued.

Populus monilifera
——— tremula
Pyrus Aria
——— Aucuparia
——— Malus floribunda
——— spectabilis
Quercus Ilex
——— Mirebecki
——— rubra
Quercus Turneri
Robinia Pseud-acacia, and vars.
Salix alba
Tilia argentea
——— europeæ
Ulmus alata
——— glabra
——— montana
Virgilia lutea

Conifers.

Abies Amabilis
——— magnifica
——— nobilis
——— Nordmanniana
——— Pinsapo
——— Webbiana
Cedrus atlantica
——— Deodara
——— Libani
Cupressus Lawsoniana
——— macrocarpa
Juniperus chinensis
——— communis
——— Sabina
——— ——— tamariscifolia
Larix europeæ
——— Kæmpferi
——— leptolepis
Picea excelsa
Pinus austriaca
——— Cembra
——— excelsa
——— laricio
——— Pinaster
——— Strobus
——— sylvestris
——— tuberculata
Retinospora ericoides
——— filicoides
——— plumosa
——— ——— aurea
Salisburia adiantifolia
Taxus baccata, and vars.
Thuja gigantea
——— Lobbii
——— occidentalis
——— Warreana
Thujopsis borealis

GRAVELLY AND SANDY.

Hardwoods.

Alianthus glandulosa
Alnus cordata
Betula alba
Carpinus betulus
Fagus sylvatica
——— ——— purpurea
Fraxinus Ornus
Gleditschia horrida
Ilex, many vars.
Juglans cinerea
——— nigra
Magnolia acuminata
Morus nigra
Platanus occidentalis
Populus Bolleana
Quercus Robur, and vars.
——— suber
Robinia Pseud-acacia
Sambucus nigra
Tilia europeæ
Ulmus alata
——— campestris
——— montana
Virgilia lutea

GRAVELLY AND SANDY—*continued.*

Conifers.

Juniperus communis	Pinus Pinaster
——— Sabina	——— pumilio
Pinus austriaca	——— sylvestris
——— halepensis	Taxus baccata
——— laricio	Thuja gigantea

CLAY.

Carpinus betulus	Quercus Ilex
Castanea vesca	——— pannonica
Cryptomeria elegans	——— Robur
——— japonica	Thuja gigantea
Gleditschia triacanthos	——— Lobbii

IRONSTONE AND COAL.

Acer Pseudo-platanus	Larix europeæ
Betula alba	——— ——— pendula
Castanea vesca	Pinus Cembra
Cerasus Padus	——— Montana
Cupressus Lawsoniana	Quercus Robur
Fraxinus excelsior	Thuja gigantea
Juniperus communis	Ulmus montana

CHAPTER XIII

TRANSPLANTING LARGE TREES

WHERE immediate effect is required, the transplanting of large trees and shrubs will be an operation of considerable importance, and though it is fraught with both trouble and expense, the results obtained have led to an increased adoption of the system during recent years.

There is hardly a limit to the size or weight of the tree to be removed when the operator is provided with suitable appliances, and success has crowned the effort of several recent operations of this nature in our Royal and other parks. Of late years several appliances for removing large and weighty trees and shrubs have been placed on the market; amongst these one of the best is certainly that made and patented by Messrs. Faulkners Ltd. Its principal recommendations are simplicity of structure, the ease with which it may be worked, and lightness, combined with such strength that even the weightiest tree can be removed. But not only is this tree-lifter valuable where transplanting is being engaged in, for in removing large tree roots, logs of timber, blocks of stone, and heavy materials generally, it has been found most useful.

The apparatus is made somewhat in the form of a four-wheeled lorry, having a steel frame only. The frame at the back is made movable to admit of the apparatus being placed so that the tree to be moved stands in the centre of the machine. Two stout planks with guide rails are laid across the trench, and the machine is backed on to these. The machine is constructed with four iron rollers, lying along over the side frame and parallel with the frame.

Around the rollers a chain is wound, the loose end being fastened to the planking which has been placed under the ball of earth containing the roots of the tree. The rollers are worked with a specially made screw-gear, which is self-sustaining, and can be moved to draw up or lower at will.

FAULKNER'S TRANSPLANTING MACHINE.

The illustration given clearly shows the apparatus with the tree being lifted for transportation. Barron's transplanting machine is also well known, and with this much valuable work in the removal of large and heavy trees and shrubs has been accomplished.

CHAPTER XIV

THINNING PLANTATIONS

In the successful rearing of timber trees for profit there is, perhaps, no other branch of more importance than a good knowledge of the art of thinning, and, at the same time, one on which so great a diversity of opinion exists.

Thinning plantations, be they old or young, must always be subject to great modification, according to the nature of the trees and soil, or ultimate design of the plantation being operated upon, and is a matter requiring great tact, forethought, and discrimination, and this can only be acquired by long experience and by those having a good insight into the peculiarities and properties of our forest trees.

The same rule as regards thinning will not apply to, say, fir, hardwood, and mixed plantations ; and far less will it hold good in the case of an ornamental and a profitable wood. No universal rule can, however, be laid down for thinning, but general principles can be given that will be sufficient for the guidance of those who have to undertake such work.

In thinning any plantation two important points are to be borne in mind—first, cut away all diseased, dead and dying trees ; second, study the relationship of trees and soil, and act accordingly. To be more explicit, we might say that in the first case, the removal of all dead and dying trees is a necessity, and that being done, one can proceed with the disposal of the standards to be left. In the second case, by suiting the trees to the soil, at least as far as possible, great benefit results, and an instance of this that came under our own observation lately will be given as an example.

A plantation 130 acres in extent was composed of oak, larch and Spanish chestnut, placed at regular distances apart throughout the whole extent of the wood. Now, the soil was gravelly for a considerable distance down, and, consequently, fairly well suited for the chestnut, but just the reverse for the larch, which, on such a soil, generally becomes " pumped," or rotten at the core. In thinning this particular plantation, it would have been very unwise to cut away the chestnut and the oak and leave the larch, and this had happened to some extent before the proper system of management and adaptability of soil to tree was thought of. Larch may look healthy enough, and show but small indications of disease, even when growing on gravel up to twenty or thirty years of age, so that it is with difficulty that the inexperienced become aware of the pending doom that usually awaits this tree when planted on such a soil. This example is merely given to point out how carefully thinning should be gone about, and that in all cases it is a wise policy to study soil in relation to the future crop of timber before an axe is laid to the tree.

At the outset of these remarks on thinning plantations it should be distinctly borne in mind that there are two chief objects for which trees may be grown, each requiring a special mode of management, in order that the best results may be obtained.

The first is their management in a purely economical sense or with a view to profit ; the second, their management with a view to ornament. There is also a third object that is well worthy of consideration, and that is the growing of timber in one and the same wood, both for ornament and utility combined, and this is very frequently the case with home woodlands that are visible from roads and drives, with strips bounding parks or pleasure-grounds, and on small properties.

To produce ornamental trees of natural appearance is by no means difficult, as by allowing the individual specimens ample room for branch development, the desired effect is gradually brought about. Far greater difficulty, however, attends the production of the greatest quantity of the most valuable trees on a given space of ground. Here many questions of the greatest moment, on which difference of opinion exists, crop up: such as at what age thinning

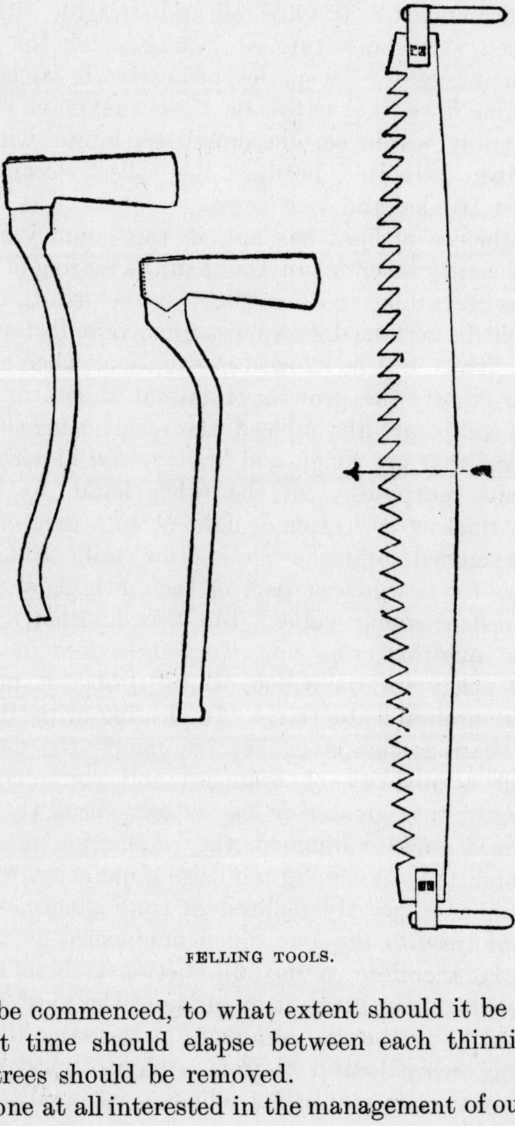

FELLING TOOLS.

should be commenced, to what extent should it be engaged in, what time should elapse between each thinning, and which trees should be removed.

Any one at all interested in the management of our woods

and plantations must have noticed that trees having an abundance of room and light on all sides make comparatively short and thick trunks that are well furnished with branches; whereas such as are grown up in a circumscribed space and amongst others are tall and straight, with clean, well-formed stems destitute of branches for the greater part of their height. All species, or nearly all, are governed by the same laws, that is to say, those that have the least room laterally within certain prescribed limits, which will be described hereafter, produce the tallest, cleanest and straightest trunks, and vice versa.

The influence of light has not, in this country at least, been sufficiently taken into account in the rearing of timber, but it has everything to do in directing the growth of trees, and should be reckoned as a most important factor by the forester. Thus, if it be allowed in excess, as when the trees stand far apart, the growth of lateral shoots and large branches will be greatly induced, the result being short and thick boles, that are rough and knotty, and ill-adapted for constructive purposes. On the other hand, by keeping the trees thick on the ground, light is, to a greater or less extent, excluded, and the trees grow tall, straight and branchless for the greater part of their height, and are of the greatest economic value. But here another and very important question crops up. To what extent in a wood, managed solely for the value of the timber it produces, will it be profitable to thin? In dealing with this, two distinct bearings should be kept in mind—the first, that too small a quantity of branches and consequently of leaves, must, to a greater or less extent, check the growth of the trees, and so diminish the production of timber; and, second, that by having too large a quantity, the value of the timber is greatly reduced in consequence, and the number of trees to the acre much diminished as well.

There is, therefore, a medium between these two, by adopting which, the greatest quantity of the most valuable timber will be produced; although, at the same time, it is astonishing, when looked at from a physiological point of view, what a small quantity of foliage is required to keep a plantation tree in a healthy, growing condition, and to produce a trunk of useful dimensions. This may, however, be explained by the fact that the almost entire absence of large limbs and branches, which in some cases would amount to one-fifth the total bulk of the tree, renders the amount of sap, and consequently of leaves, required proportionately less, the nutriment being mainly elaborated in the building of the trunk.

Independent altogether of the forests of northern Europe, Canada, etc., which have grown up naturally, examples might be given in our own country—the native Fir forests in the Highlands of Scotland, the valuable Beech woods on the Chiltern Hills, and a few Larch plantations in Yorkshire and Bedfordshire—where trees averaging 70 ft. in height, growing at from 4 to 6 yards apart, and with only a tuft of foliage atop, may be seen. It is only, however, by long and careful study and attention to different trees at different stages of their growth that any idea can be formed of the amount of branches and foliage required for preparing the sap that will be necessary for the profitable yearly increase of the trunk until maturity is arrived at.

The questions now to be dealt with in growing timber for purely economical purposes are: (1) At what age should thinning be commenced? (2) To what extent should it be engaged in? (3) What time should elapse between each thinning? And (4) Which trees should be removed?

At the outset, it may be well to mention, that in so far as any of the above questions are concerned, no hard and fast line can be laid down as to the universal management of plantations, each tract of wood, and even different parts of the same wood, according to the particular species of tree, quality of soil, altitude, and exposure, being dealt with on its own peculiar footing. Many serious mistakes have been committed, and irreparable damage done, by treating every plantation alike and according to the rule-of-thumb method. Upon the manner in which thinning is performed, much of the ultimate success of a plantation will depend, and it is a matter requiring great tact, forethought, and discrimination, which can only be acquired by long experience, and an intimate knowledge of trees and their surroundings.

The quality as well as quantity of timber produced should be carefully considered in the management of every plantation. In dealing with the following questions regarding thinning, it may be well to state that the average plantation will alone form the basis of remarks.

1. **At what age should thinning be commenced ?**—The necessity of thinning, few practical arboriculturists will care to deny, and that whether the plantation be composed of coniferous or hard-wooded trees. It is the abuse of the practice that, in this country at least, is so to be condemned, and I entirely dissent from those who consider that a coniferous plantation should be left to Nature, or, in other words, should thin itself. The argument that the natural forests of America, the source from which we derive the finest and best quality of timber, thinned themselves, is often quoted ; but it should also be remembered that the conditions of such forests are widely different from those of our own, and that the waste of timber is immense, great quantities being destroyed in procuring what is required. Although, however, we cannot adopt all the details of Nature's practice, we can and we should admit the correctness of the principle on which she acts, and by studying this we learn much, and it is by acquiring a knowledge of her economy, and where and when to apply assistance, that the point of the whole argument rests. Natural regeneration, unless it be in a limited number of cases, and particularly with our least valuable classes of timber, is never, in this country, likely to supersede artificial planting, and it has been conclusively proved in the Forest of Dean, that with our most valuable timber-producing tree, the Oak, the latter system is most to be encouraged.

No thinning of young trees should take place until a complete ground shade has been established ; and it is most important for the welfare of the plantation, that a complete overhead foliage covering be brought about at as early a date as possible after planting. At what age this may take place will depend mainly on the size of the plants used and the distance apart at which they were inserted in the ground, and to a lesser extent on the quality of soil and other considerations. At a short period after a complete leaf canopy has been established, the individual trees begin to press against each other, and later on a struggle for existence commences, the stronger specimens gaining the supremacy over the weaker. Here it must be borne in mind that hard-wooded trees require proportionately more room for their healthy development than coniferous kinds, while length and clearness of stem, produced by a due proportion of shade, is an object of prime importance. It might, as some suppose, do little harm, to let the struggle for existence go on unchecked, but there can be no question that for several reasons it is wise policy to allow the stronger trees every chance of succeeding, and to cut away the weaker. The object should be to provide for the trees left standing that amount of room or growing space best suited for bringing about the particular conditions aimed at, and in thinning, it must be the aim of the forester to arrive at the happy mean—neither over nor underdoing the work, the former in particular. But while overcrowding is not to be tolerated, the danger of suddenly exposing the trees to currents of air, to which they have hitherto been unaccustomed, must be carefully guarded against, and it should be distinctly borne in mind that to thin trees in an abrupt manner is one of the greatest mistakes that it is possible to make.

Statistics compiled from a healthy Larch plantation, growing on fair soil, and in a moderately sheltered position in southern England, will now be given, from which a good idea can be formed as to the age and size when thinning should be commenced. The trees when planted were 2 ft. high, and pitted at 3 ft. apart. In four years the outer branches began to touch each other, and in six years from time of planting the average height of trees was 7 ft. 10 in. and the shade occasioned had killed out most of the grassy undergrowth. At this period of growth, the disproportion in the size of the trees was, as is usually the case, considerable, and left no doubt about which to remove when the first thinning took place. Two years afterwards, or in eight years from time of planting, thinning was first engaged in, the taller trees at that time averaging nearly 11 ft. in height, but many of the suppressed and weakly were little more than half that size. Another thinning took place during the twelfth year, and in 1911 the trees average 40 ft. in height and were standing at a distance of 9 ft. apart. The plantation referred to is on the Earl of Derby's property of Holwood, in Kent.

2. **To what extent should thinning be engaged in ?**—In following up the latter case, the tallest and healthiest trees were reserved; all distorted, sickly, and stunted specimens being removed; but in the case of two or more proportionately small trees growing in close proximity, the most promising was left, and the others cut away, thus avoiding gaps in the plantation. At the first thinning it will be found quite impossible to leave the trees anything like regular over the ground, although this should be studied as much as possible, and bare spaces are to be carefully guarded against. With the vigorous growth of the trees, and the interlacing of branches previous to the first thinning, no great openings will require to be made, and nothing more than will be canopied over during the next two years.

3. **What time should elapse between each thinning ?**—This can only be correctly decided after a careful inspection of the particular wood. Generally speaking, after the first thinning, when the trees were eight years old, the lower branches gradually began to give way, and as the trees increased in height, this became more and more apparent, and formed a good guide as to the time which should elapse before the second thinning might profitably be engaged in. The intervals between the various thinnings should for the first thirty years in the case of Larch—and, indeed, most other trees—be comparatively short, but become longer with advance of age; but heavy thinnings must be carefully avoided, especially if the best class of coniferous timber is to be produced.

A well-managed Larch plantation of twenty years'

growth should have the trees branchless for about one-half of their height, which is, of course, brought about by crowding, and at forty years three-fourths of the trunk should be clear of branches.

I have purposely abstained from giving the number of feet apart at which trees ought to stand at various stages of their growth, and the number of thinnings they require, as being likely to prove misleading.

4. **The question as to which trees should be removed** need cause little or no anxiety, as the dead, dying, diseased, and badly-formed will naturally receive first attention.

In thinning, the following short rules should be observed :—

1. Thin not at all until the undergrowing vegetation has been completely killed out by the overhead foliage, the golden rule of sylviculture being to keep the sunshine off the ground until near the end of the tree's career.

2. Thinning should be performed in such a manner as not to reduce the value of the crop, but so as to tend to the production of the greatest quantity of the most valuable timber in the shortest possible space of time.

3. By thinning allow of sufficient light and air to prevent the trees becoming drawn up and lanky, but avoid too much space, which induces the growth of side branches and detracts from the value of the timber.

4. Keep up the number of trees to the highest possible pitch until they are tall, straight and clean, and thin gradually.

5. The danger of suddenly exposing the trees to currents of cold air to which they have hitherto been unaccustomed must be carefully guarded against and it is a serious mistake to thin in an abrupt or unequal manner.

6. In thinning remove all dead and dying trees first, then the distorted and stunted, the best grown and healthiest being preserved for the permanent crop.

7. Hard-wooded trees require proportionately more space for development than conifers, and the annual rings in the timber of the latter should be narrow in proportion to those of the former, good quality of hard-wooded timber being

indicated by broad annual rings, but with coniferous wood the reverse is the case.

8. Thinning should be commenced at the central or most sheltered point of a plantation, or section of same, so that the outer intact boundary may continue to form a barrier to cold winds, which might prove injurious to trees that had hitherto stood in close order.

9. In economic forestry, timber trees should be cut down when they have arrived at maturity or have ceased to grow. With coniferous trees this generally occurs at from seventy to ninety years; but with hardwoods, excepting the ash and chestnut, which are of most value, say, up to fifty years' growth, the period may be greatly extended.

10. Thinning is usually performed in autumn and early winter, but it may be well to remember that at the latter period the lowest percentage of moisture (about 47 per cent.) is present, and the timber consequently of the greatest value for constructive purposes.

CHAPTER XV

TREE-PRUNING IN ECONOMIC FORESTRY

WHERE trees are grown for profit, they will, if properly managed, prune themselves, and where for ornament the natural outline is far better than any of the contortions and symmetrical shapes that have been recommended by various writers on the subject.

A broken or dead branch may be removed, a rival leading shoot cut away, or an ungainly or dangerous limb amputated, but here all pruning should cease, the practice being wholly wrong and unreasonable, and without one recommendation to be adduced in its favour. In an economic way the finest plantations of either coniferous or hard-wooded trees in this country are those where the individual specimens are growing so thickly together that the branches are killed outright for fully one-half of their height. Here the stems will be straight and clean, and the timber when converted free from the knots and warping that are so characteristic either of standard specimens or such as have been grown too thinly on the ground.

Every one knows that an Oak growing alone or along the margins of a wood is in nine cases out of ten branched almost to the ground, and the bole in consequence rough and ill-fitted for any particular constructive purpose, and the same may be said of every other tree, be it hard-wooded or coniferous. Larch and Scotch Fir trees growing along the margins of plantations are rough and knotty, and sell at a considerably lower figure compared with those further in, where the branches have been killed back gradually as the trees increased in size.

The same thing is markedly the case in young woods of

ash, oak and chestnut, where they have been grown sufficiently thick on the ground to kill off the lower branches, and also to cause the trees to rise straight, clean and tapering. It is a well-known fact, too, that the timber of trees so grown is far more elastic and realizes a much higher price than that of the same age grown under conditions where pruning might have been a necessity. A case of this kind came under my own notice only a short time ago in which one-half of a plantation of hard-wooded trees realized fully one-fourth more than the remaining half. It came about in this way. Both ends and a large patch in the centre of the wood had been thinned out severely for the purpose of planting game covert. The trees, standing thinly on the ground, branched out and soon covered the open spaces where underwood had been planted. In thinning the whole plantation the trees on these particular parts were very rough and knotty, and bore no comparison to those where they had been left moderately thick on the ground, in consequence of which the boles were straight, clean and tapering. This case is specially noteworthy, inasmuch as the trees over the whole area were growing under exactly similar conditions as to soil, shelter, etc., and were of the same age and species.

Great and irreparable damage has been done to woods and plantations in this country by too heavy thinnings, by commencing the thinnings at too early a period, and by adopting the book method of leaving the trees at measured distances apart and a stated number to the acre according to the age of the plantation. Such rules can never be expected to work satisfactorily, the size of trees depending so much on the character of the soil, exposure of the woodland, and other peculiarities of the particular district in which they are planted.

Timely and judicious thinning should never be neglected, but it is the over-thinning, whereby branches and knotty trunks are produced and the supposed need for pruning follows, that I wish to deprecate and entirely dissent from. Grow your timber trees so thickly on the ground that the stems are induced to become straight, clean and branch-

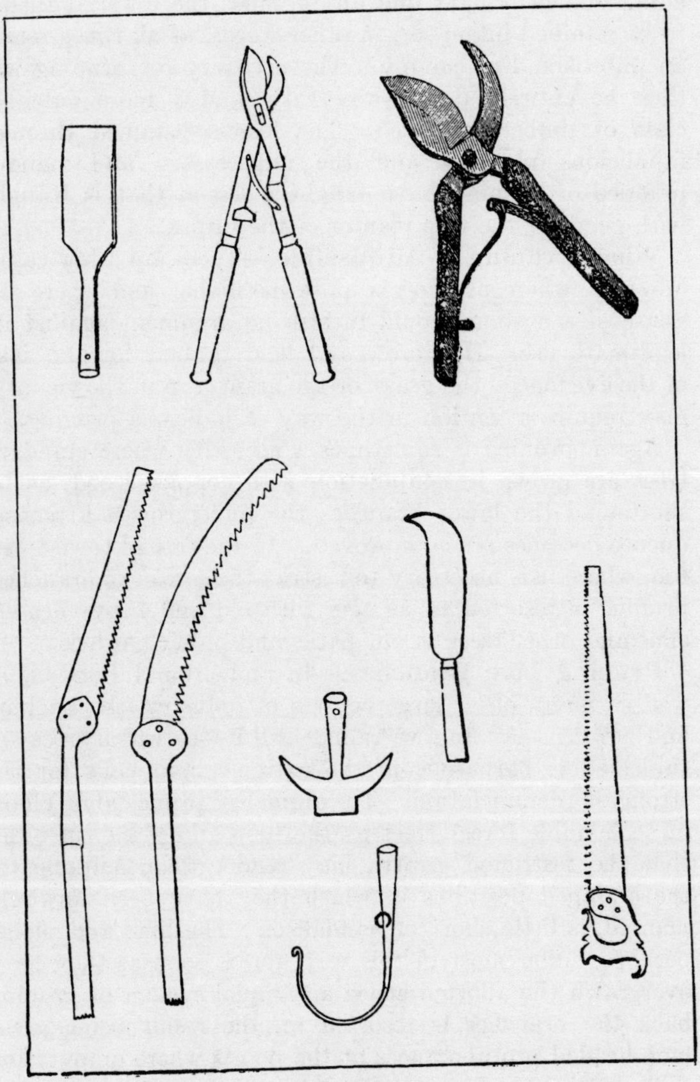

PRUNING TOOLS

less for the greater part of their height, and on no account admit sufficient light and air to cause the lower branches to be retained intact, or, in other words, at all times retain an unbroken leaf canopy. The necessity for pruning will then be entirely done away with, and a more valuable class of timber produced. The losses sustained through injudicious planting and the unnecessary and ruinous practice of pruning have taught a lesson that is fraught with good for the tree planter of the future.

When Pruning is Admissible.—There are a few cases, however, where pruning is quite justifiable, and where the abuse of a system should furnish no argument against its legitimate use. Hedgerow and field timber, for the sake of the live fences, the grass, or the grain crop in the vicinity, may require attention in the way of judicious pruning.

Again, pruning is sometimes a necessity where standard trees are grown in conjunction with coppice wood, as by shortening the lower branches the undergrowth in consequence becomes much improved. In the case of town trees, too, where it is necessary to restrict the spread of branches, pruning is resorted to, as also with old and heavy-headed elms and other trees in our parks and public gardens.

Pruning Live Branches.—In and around London, as well as many other large centres of industry, the hacking and hewing—pruning we cannot call it—to which trees are subjected is barbarous in the extreme, and calls for the strongest denunciation. To annually prune and elbow in such noble forest trees as the lime and plane, in order that the restricted growth may render them suitable for the cramped positions in which they have been unwisely planted, is little short of vandalism. The lime and plane, perhaps, suffer most in this way, for as soon as they have overgrown the allotted space an annual system of pruning back the branches is resorted to, the result being great mop-headed protuberances at the points where amputation took place, which not only rob the tree of its graceful natural appearance, but render it susceptible to disease and insect pests. There is no need to specialize cases where this most objectionable system is carried out, for a walk around our squares and gardens will unfortunately reveal how prevalent is the maltreatment of trees in the way of pruning. There might be some excuse for planting our noblest forest trees in cramped and unsuitable positions were there no other species of smaller growth that would take their place, but the Pyrus, Cratægus, dwarf Acacia, Mulberry, Catalpa and Sumach surely give us sufficient scope for choosing trees of restricted growth for confined positions and so do away with the barbarous system of pruning which the use of large-growing species necessitates.

BAD EFFECTS OF PRUNING

Even in the case of dwarf avenues and screens it is quite unnecessary to use such large-growing trees as the lime and plane where pruning must oft be resorted to, for an avenue or screen of thorn or mulberry, the beam tree or mountain ash would be more suitable, and in the end far more natural and artistic in appearance.

To sum up in a few words, my contention is that no tree should be planted in a position where, in order to keep it within due bounds, a systematic clipping and pruning has to be resorted to.

Pruning Dead Wood.—Opinions differ greatly as to whether or not dead branches should be removed from park and woodland trees. It is, however, mainly a matter of taste, and a point on which two of the largest owners of woodlands in this country hold distinctly opposite opinions, though at the same time it cannot be denied that the careful removal of all dead and dying wood from a tree is highly beneficial. Pruning should, however, only be extended to such trees as are in a fairly healthy condition, with well-developed heads, and containing only a moderate quantity of dead wood, there being many fine old specimens that would be rendered unsightly in the extreme and receive no benefit from removal of the dead and dying timber, but this has direct reference to trees standing singly throughout the park, and not to specimens in the woodland.

In many of our parks and woodlands at the present time there exists an undue quantity of dead and dying wood, which may be attributed to natural decay, the quality of the soil, and in some instances to long-standing neglect in the matter of non-attention to wounds which have been caused by wind-broken limbs and branches. Such trees would be greatly improved, both in health and appearance, by judicious removal of the dead branches and attention to old wounds in order to prevent the ingress of water, the decay of many branches being directly attributable to this cause. That an undue quantity of dead wood will induce injurious insect pests, such as the goat and wood leopard moths, which attack healthy trees, is well known, and was exemplified in one of our London parks recently where numerous young thorns and various species of Pyrus were injured by the latter. In removing dead branches cut them well back into the living wood in order to induce fresh growth at the point where amputation takes place. The removal of large dead limbs from old specimen trees is an operation that requires a great amount of skill, and should only be entrusted to those who have had practical experience of pruning in its various phases. As before stated, the removal of dead wood is merely a matter of sentiment, though of its practical utility there can be no doubt, and in all cases where the stag-headed trees are conspicuous, the removal of the dead wood is to be recommended.

Pruning Shrubs.—Generally speaking, shrubs are pruned with little or no consideration as to whether they will be benefited by the operation. While symmetry and regularity of outline are to be admired in a shrub, these qualities should never be gained at the expense of natural grace and production of flowers. The judicious pruner will, therefore, aim at preserving the peculiar habit of each shrub as far as possible, while interfering but little with the production of flowers. The various species of Deutzia, Forsythia, Philadelphus and Weigela flower on the wood of the previous year's growth ; therefore such shrubs should be pruned immediately after the flowering season—say in June, but never in spring or winter—at least, if the production of flowers is to be taken into account. Again, the various species of Syringa, Spiræa, Lonicera and Hibiscus may safely be pruned during winter, the flowers being produced on the young wood ; while *Hydrangea paniculata grandiflora* must be severely pruned in early spring, for only by so doing will the greatest wealth of flowers be produced. *Chimonanthus fragrans* should be pruned in February ; while the various species of Ceanothus should not be touched till all danger of frost is past. *Kerria japonica* should be pruned in autumn, when old wood may be cut away.

All pruning operations should be carefully carried out with a sharp knife and not with the pruning shears, the point of amputation being always close to an eye or bud. Too severe pruning should be avoided, a judicious thinning out of the branches being far preferable to indiscriminate shearing and cutting back.

How and when to Prune.—The latter part of May or beginning of June is undoubtedly the best season for pruning the majority of hard-wooded trees, as during that time the motion of the sap is most vigorous, and in consequence the wounds caused by amputation heal most quickly.

There are a few exceptions—the birch, sycamore and maple—where, on account of profuse bleeding, pruning had best be postponed till after full foliage has been attained.

Much mischief has been done by the pruning knife, and still more by the handbill and saw when placed in the hands of inexperienced workmen—a fact that is apparent to any interested person who visits various parts of the country.

When conducted with care, on sound principles, the effect of pruning on the class of timber referred to is highly beneficial, whereas, when done in a haphazard way and by an inexperienced hand, it is often attended with great danger and grave results.

In cutting or foreshortening small branches, a sharp pocket-knife will be found most convenient, but when large limbs have to be removed the pruning saw should be brought into request, The branch to be removed, especially if of large size and weighty. should first be cut through at any convenient distance from the main stem, thus preventing splitting and tearing of the bark, after which the stump may be neatly sawn through as close to the bole of the tree as possible. Undercutting by a few draughts of the saw will here also go far in preventing tearing of the wood and bark. So as to render the saw-cut smooth and prevent water lodging on the surface, the face and edge should be neatly dressed with an adze or sharp pruning knife, and then painted with tar. When cutting over an upright-growing branch, such as in pollarding trees, etc., never cut on the horizontal, but always in a sloping direction, so that the rain may pass off quickly.

CHAPTER XVI

BRACING AND REPAIRING TREES

THOUGH occasionally resorted to in olden times, particularly in the case of orchard trees, it is only of late years that the systematic treatment of tree wounds and diseases has been generally adopted. Even at the present time, the filling of hollow trunks, bracing of heavy and diseased branches and attention to insect and fungoid pests are rarely engaged in and but imperfectly understood.

When the value of old trees in an ornamental or utilitarian sense is taken into account, it is remarkable what a small amount of attention they receive when subjected to accident or disease. As a general rule, cavities or hollows in a tree stem, if left to themselves, gradually increase in size until the ascending sap is entirely cut off, when the crumbling stem either falls to pieces or is broken over by the wind. A branch requires to be braced or strengthened when from its weight and shape it is likely to get wrenched from the main stem during stormy weather and so injure the tree or mar its natural beauty.

Both insect and fungoid pests do a considerable amount of damage to trees, both young and old, but as special chapters are devoted to their depredations they need only receive a passing notice here.

Hollow Trunks and their Treatment.—However desirable it may be to repair hollow trunks, yet a great deal of discrimination is necessary in deciding which trees should be operated upon, especially in the case of such as are reduced to mere shells and are not likely to derive any benefit commensurate with the labour and cost of materials involved in bracing and filling.

Young, healthy, vigorous growing trees, that have become damaged or diseased, should in the majority of instances be attended to ; but in the case of old specimens on which a great amount of labour and money would necessarily have to be expended, several important bearings must be considered before works of repair are taken in hand. Historic and valuable trees, those occupying prominent positions on a lawn or park, town trees where vegetation is scarce, and such as are not too old and fragile, may be dealt with within certain bounds, but to doctor all diseased and hollow trees on the most approved principle in such places as Epping Forest or Burnham Beeches would be highly imprudent, whether in view of the little benefit which in many instances would accrue by so doing or of the great expense involved in such an operation.

The filling, too, must be adapted to circumstances, and to deal with the entire trunk of a hollow tree is in most cases quite out of the question, both from the point of utility and expense. Short-lived trees, such as the Alder, Birch and Poplar, or such as are liable to sudden attacks of insect and fungoid pests, and those that are unfavourably situated in mining and chemical areas, should be scantily dealt with, but the practical woodman will see at a glance which trees are best worthy of his attention and treatment.

Probably the simplest, and certainly the cheapest method of dealing with diseased and hollow trunks is to clean out thoroughly all dead and decaying matter, the interior being scraped and swept with a rough brush, so that the loose rotting wood and bark is removed. When quite dry, the interior of the trunk should be painted with one, or preferably with two coats of creosote or carbolineum, and filled up with a composition of one part of Portland cement to three of clean gravel and sand, the surface coating at the orifice containing the largest quantity of cement. A coat of coal tar on top of the dry creosote goes far in making the concrete adhere firmly to the wood.

When a fairly healthy tree is being operated upon, and when the bark is likely to grow over the exposed surface, the concrete should only be brought up on a level with the underside of the living bark; in other words, the cambium should be left free for expansion. Sometimes, brickbats, broken small, are used instead of gravel, but for various reasons concrete is preferable. The concrete surface may be prevented from cracking by applying annually a coat of paint, which, for appearance sake, may be of a similar colour to that of the bark of the tree.

In the case of a large tree in which the trunk is quite hollow, usually with a basal and top opening, the amount of cement required to fill the cavity precludes the possibility of it being employed on the score of economy, but where the hollow only extends for a comparatively short distance upwards and inwards the composition is to be recommended. When a large amount of material is required to fill a hollow stem, clean bricks broken to about the size of a golf ball may be used, but they must be packed firmly and the surface, wherever it comes in contact with the weather, glazed over with cement in order to effectually shut out dampness. Asphalt has of late years been employed in the filling of hollow stems, and is to be recommended on account of its elasticity and lasting properties.

Sometimes, the interior of a large, hollow tree stem is strengthened by crossbeams of timber, any holes which extend to the outside being carefully covered with sheet lead or zinc so as to prevent the ingress of water. A sheet of lead or zinc is, however, only a superficial remedy and should be dispensed with where filling the cavity is at all practicable. Lead, in some cases, is preferable to zinc for covering wounds in trees, as it fits into position more readily and is easily fixed.

The treatment of hollow stems without filling the cavity is in certain instances quite permissible, and can be carried out at a comparatively small cost. Broadly speaking, this method consists in cleaning out all decayed and decaying matter, painting the interior with carbolineum or tar, and, where possible, bracing the trunk from the interior by means of stout crossbars of wood and covering surface holes with lead, zinc or sheet copper. When a tree trunk is hollow from top to bottom this method is to be recommended, and stout struts placed within the cavity at various of the weaker points will go far towards preserving many an aged specimen. When the cavity in a hollow stem is so large that a person can enter it, the work of either filling with a suitable composition or supporting with wooden struts is greatly simplified.

Recent experiments have proved the value of a mixture of sawdust and asphalt for filling cavities. It is particularly valuable in cases where concrete is too rigid and unyielding, such as in dealing with trunks and branches that are apt to be swayed about in stormy weather. The materials are dry sawdust, that of Oak, Chestnut and Beech being preferable, and solid asphalt derived from the refining of petroleum, which is at present in use for filling the interstices of street pavements in various parts of the metropolis.

For filling cavities in heavy, swaying branches, one part of asphalt to four of clean, dry sawdust will form a mixture that is non-rigid and yields with the motion of the branch in which it is inserted. In dealing with the trunk, which is more rigid and less affected in stormy weather, a larger quantity of sawdust should be used. The filling is made by stirring dry sawdust into boiling asphalt until the desired consistency is reached, and before the composition has cooled, it should be inserted in the previously prepared cavity. As in cement filling, the preparation of cavities to be filled with sawdust and asphalt should be carefully attended to, the decayed wood and all soft and rotten material being removed, and the interior surface rendered sterile by an application of carbolineum or kerosene. For wound dressings, gas tar and liquid asphalt is to be recommended, this combination forming a more continuous and elastic covering than is the case when tar or paint alone is used.

Where the cavity is small and, as is often the case, filled with water and decayed and decaying vegetable matter, the following course is recommended :—When only a few inches deep, the water can usually be got rid of by mopping it out with a sponge attached to a convenient handle, but when the hole is too deep for this method the water may be extracted by tapping, that is by boring a hole with an auger through the trunk to the bottom of the hole. A half-inch auger will suffice, and the correct spot to bore can usually be ascertained by measuring the depth of the hole. After the water has been removed the cavity must be thoroughly cleared of all decaying tissue and accumulated vegetable matter, and in order that this may be done effectually, it may sometimes be necessary to enlarge the opening so that a suitable tool can be inserted.

When the cavity has become quite dry—a matter of considerable importance—it should be treated with creosote or carbolineum and afterwards filled with concrete in the proportion of one part of cement to four of clean gravel. Ram the concrete firm so that the cavity is perfectly filled, and finish off at the orifice with pure cement in such a way that the cambium may be induced to form a growth over the edge of the filling.

A cheap but temporary method of dealing with small hollows and cavities in fruit and other trees, is to fill these with a mixture of clay and cow-dung in the proportion of two of the former to one of the latter. Knead well and apply when in the consistency of putty, cleaning and disinfecting the hollows before filling.

Supporting Heavy and Diseased Branches.—A limb requires to be braced or strengthened when from its weight and shape it is likely to get wrenched from the main stem during stormy weather, when the weight of the minor branches and foliage is too great for the strength of the limb, when the wood of the tree is unusually brittle, when trunk or limb is decayed, and when, from accident, the tree has become one-sided and lost its natural appearance.

Forked trees often require staying, as also do such as have suddenly become exposed to storms to which they have hitherto been unaccustomed. But probably in the matter of bracing and strengthening, the most important of all trees are those in our public parks and such as are contiguous to or overshadow dwelling-houses. Park trees should receive special attention in the way of making sure that limbs are secure and not likely to cause injury to visitors; while in the case of heavy branches hanging over dwelling-houses these should be carefully and periodically examined and made secure whenever it is found necessary. Should there be any doubt as to whether a limb is insecure and dangerous, a decision should be made in favour of bracing or reducing the weight by pruning.

Unwieldy and heavy limbs are most commonly to be found on isolated trees, or such as have had plenty of room for the perfect development of stem and branch. As the loss of one or more of the larger branches often mars the ornamental appearance of a specimen tree, every reasonable means should be employed to prevent such a disfigurement. By reducing the weight of a branch by pruning, or by giving support to any that are heavy and diseased by means of wires, chains or light iron bands and connecting rods, the trouble may be averted. In all cases the object should be to unite the branches in such a way that they may offer the greatest amount of resistance to the storm, and at the same time be neither conspicuous nor clumsy in appearance owing to the operation.

Chains, though often used, probably owing to their being readily obtainable and cheap, are for various reasons to be avoided, the flat iron band lined with leather or rubber being preferable, more readily adjusted, and less likely to cut into the bark and wood. The bands, which can be made by any blacksmith, are usually 2½ in. wide, and of the shape of the branch to be encircled, each being in two parts to facilitate fixing and to allow of slackening at any future time should the necessity arise. The band is made of a larger size than the branch to be encircled, so as to allow of the insertion of a leather or rubber collar between it and the wood, the purpose of the packing being to prevent undue friction and chafing of the bark owing to the movement of the tree. The bands, whether placed around two opposite branches or the main stem and a branch, are connected together by a light iron rod and, according to the strain, may be from one-half to three-quarters of an inch in diameter. This rod, like the bands, is divided into two parts, which are connected by a swivel and screw for convenience in loosening or tightening.

BRACING A TREE

Great care is necessary in choosing the point at which a band should be placed, and in deciding which branches or branch and stem should be joined together, so that the greatest resistance may be obtained and in order that the one may act as a support to the other. The shape of the tree and disposition of the stem and branches can alone be the guides in this matter. In comparison with the use of a chain, the advantages of this method of supporting heavy branches are principally ease of fixing, greater rigidity, less friction, adaptability for loosening or tightening as may be required, and better results.

Accurate measurements as to the size and shape of the limb to be operated upon must be taken and the girdling hoops made accordingly. The best way to take these is to ascend the tree, and, after deciding as to the points where bracing is to take place, encircle each branch with a strip of hoop iron, which will not only give the size but shape of the particular limb, a most important point where neatness is studied and after-chafing of the bark and wood is to be avoided.

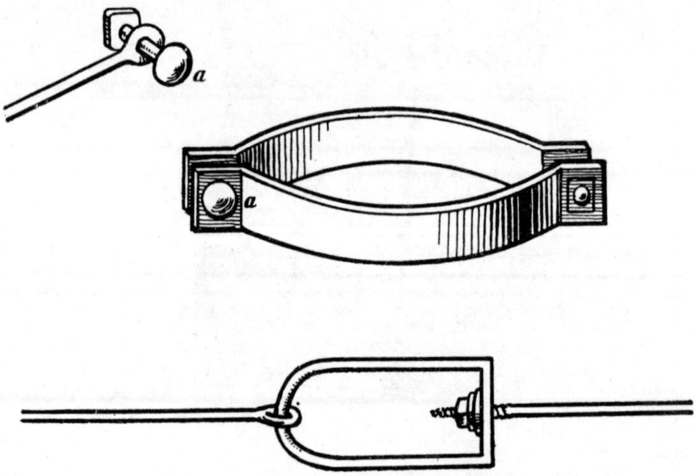

BAND AND CONNECTING-ROD

In the case of old trees, or such as are not likely to greatly increase in size, the bands need not be much larger than the actual size of the branches to which they are to be affixed, but where the tree is young and vigorous, room for expansion should be provided. Fixing the bands and connecting-rods is readily accomplished either by the use of ladders or by climbing the trunk and slinging the portions into position by means of ropes. The exigencies of each case will point out the best means of carrying out the work, as also the height at which the supports can most effectively be placed. When damaged or diseased branches are being dealt with, great care is necessary to ensure that the bands are placed in such positions that the greatest leverage against wind is afforded, and so as to minimize the risk of the branch being broken across at the weakened point during stormy weather.

Frequently, where the main trunk divides into two or more portions near ground level, it will be found that a split or crack has been caused by the swaying of the divided stems. In such cases the split between the stems should be carefully cleaned out, treated with an antiseptic and filled with asphalt. The limbs should then be braced together by means of a band and connecting-rod, and, as the asphalt becomes set, the brace may be tightened up as may be found necessary.

Rarely is it found necessary to place a band around the stem of a tree to prevent splitting. Where, however, appearances point to the likelihood of this taking place, as in the case of forked trees, especially when the stem divides abruptly into two or three heavy limbs, it is a wise precaution to encircle the bole with a wide band of iron. This band should be placed at such a height that the greatest possible resistance to splitting of the stem is afforded. Such bands are usually proportionate to the strain of the stem to which they are to act as support. A band of leather or rubber between the iron and bark of the tree to allow for expansion of the stem is necessary.

Supporting tree branches by chains is not to be recommended, for the simple reason that, however carefully they may be adjusted, friction and chafing of the bark and wood is sooner or later bound to ensue. Connecting the flat iron bands, already referred to, by means of chains or strand wire, is, however, not so open to objection nor attended with such ill effects as is the case when the chain or wire is placed in direct contact with the branch, for even with the most careful adjusting and packing, the uneven surface of a chain, with its saw-like action during stormy weather, soon renders the belting useless for the purpose intended, and both bark and wood suffer in consequence.

Occasionally we see holes bored through the live branch or trunk in order to fix the binding rod or chain—a most reprehensible practice that, fortunately, has few supporters amongst those who have even had a cursory knowledge of trees and their growth.

It is, however, generally admitted that there are several disadvantages connected with bracing trees with iron rods that pass through holes bored in the trunk or branch. However carefully this operation may be performed, there are grave risks from insect and fungoid attacks in holes that have been made in living wood, as the friction caused by the movement of the tree renders healing of the wound quite problematical. Further, the movement of an embedded iron rod is not only apt to cause friction with the wood but the strain must often be in an opposite direction to that which was intended, this increasing with the diameter of the stem or branch.

Another untidy, slipshod and most objectionable method of staying a tree is by using wire as a band around the trunk. This certainly has the virtue of cheapness, but in so far as efficiency is concerned, and especially in the case of old trees, it is to be deprecated.

In the repairing and strengthening of tree limbs it should consistently be borne in mind that any girdle or band that either partially or wholly constricts the growth of a branch defeats its own object by preventing the uninterrupted flow of sap and impairing the strength of the limb. With care in forming the band and a proper method of adjusting it, there is no reason why the desired object of strengthening a stem or branch may not be attained. On the other hand, a badly formed, narrow band, too tightly applied and without the requisite packing, may in a short space of time be productive of the most unsatisfactory results, particularly in the case of young and fast-growing trees. The main point is that the bands should be fixed in such a way that friction to the bark is avoided and the flow of sap uninterrupted. Unfortunately in the past this has not in all cases been carefully provided against, with the result that the bands, owing to their being too small, gradually became embedded in the wood and, by injuring the cambium and preventing the flow of sap, defeated the object for which they were intended. For fuller information on doctoring trees, the reader is referred to my book on *Tree Wounds and Diseases*.

CHAPTER XVII

COPPICE AND UNDERWOOD : THE MANAGEMENT OF COPPICE WOOD

ALTHOUGH coppice wood has deteriorated considerably in value during late years, yet in certain districts and on suitable soils its cultivation is still to be recommended. Unfortunately, however, the uses to which coppice wood are in the main applied are much localized, and this, with the bulky nature of the commodity, will to a very considerable extent restrict its sale to the localities in which it is used. Thus in parts of Kent and Sussex hop poles find a ready market, pea and bean stakes in the gardening districts around London, crate wood in the neighbourhood of potteries, willow and ash in basket-making localities, and faggot wood wherever kilns abound, or in the environs of populous districts generally. It will, therefore, be seen that next to the question of soil probably the most important is that of local demand, so that in forming a coppice plantation only such underwood should be used as there is a demand for in the particular district. Foreign importations and preferential carriage rates have also dealt hardly with the profitable cultivation of coppice wood. Coppice wood may either be grown alone or in company with large standard trees, but the latter must at all times be kept sufficiently thin, so as not to overshadow and kill out the undergrowth. There are advantages, too, in employing standards for the protection they afford to the young shoots in spring, as also in the amount realized for the periodical thinnings to which they may be subjected. For this purpose the oak is to be recommended, but such wide-spreading trees as the ash, elm and beech, which produce so dense a shade as to kill out or seriously injure all vegetation that might spring up beneath them, are to be avoided. Generally where the health and vigour of the coppice wood are points of first consideration, it will not be advisable to allow the standard trees to occupy altogether more than about one-fifth of the wooded area; and even then the lower branches should be pruned off, so that the effects of shade will be mitigated as much as possible. As to the woods which coppice most freely, the ash, oak and hazel occupy the first rank, at least in a profitable sense, the elm, willow, beech, birch, hornbeam, alder and sycamore occupying a second, but, as before stated, the nature of the soil, and less so the altitude and exposure, have everything to do with the particular species that will succeed best. Thus ash will do well where the soil is moist and loamy, the Spanish chestnut in sandy or gravelly districts, for rich plains and hollows the oak will be most remunerative, the alder and willow in marshy ground, and where bare and exposed, the birch, hazel, beech and hornbeam will succeed best.

The preparation of the land and planting for coppice are similar in all respects to that adopted for the growing of an ordinary timber crop. Where the ground is too wet, draining should be judiciously engaged in, while trenching, although expensive at first, is amply compensated for in the increased growth and vigour of the underwood. The pits for planting may be made from 3½ ft. to 4 ft. apart, and, if the ground was previously trenched, of sufficient size to hold the roots without undue cramping. It is always well to keep the stools tolerably close together, as the shoots take a more erect habit and are straighter and more valuable than when allowed too much space and side room. Two years after being planted, or at the end of the second autumn, the young trees, excepting such as it may be thought advisable to leave as standards, should be cut over near ground level. This cutting is a most important operation, and should only be performed by skilled workmen, with tools of the best description well sharpened. The cut should be clean and directed upwards, all splitting of the stems and tearing of the bark being assiduously guarded against as conducive to decay and early death to the stools. After four years' growth the shoots should be thinned out, leaving, say, four on each stool, and these preferably the strongest, the work being carried out at any time from November to the end of March, but not during frosty weather.

Upon the kind of wood grown and uses for which it is designed will depend very much the length of rotation pursued, for while osiers might profitably be cut at the end of the second year, ash, oak and chestnut would not usually, even on the best quality of soil, be felled sooner than from ten to twelve years, and the poorer classes of coppice wood, especially on light soil, at from twelve to sixteen years. It should be remembered, however, that the duration of the stool is usually proportionate to the length of the rotation adopted, and with good management on fair soil the best class of coppice wood has a duration of nearly a century. In felling the coppice wood it is always advisable to cut as near the ground level as possible, the shoots sent up having the advantage of rooting in the ground and so extending the area of the stool. The system of allowing the stools, by careless cutting, to rise in some instances several feet from the ground is contrary to the methods adopted under the best management. A sharp billhook should be used for all smaller shoots, a light well-ground axe for those from 3 to, say, 6 inches in diameter, and the cross-cut saw for all over that size.

The coppice wood is usually sorted out after being cut down, the best poles being laid aside for the use of the hop grower, the next size for pit props, or fencing according to the demand of the moment, and so on until every pole has been sorted according to the use for which it may be intended, the lop and branches being bound into faggots for fire or oven-lighting.

The following are the various uses to which underwood is applied: hoops, hurdles, crate rods, pea stakes, spars for thatching, withes for faggot-tying, sheep cages, hop poles, brooms, broom handles, skewers for butchers, chisel handles, plant stakes, whip handles, gunpowder wood and faggots.

Profits will vary from 15*s.* to 25*s.* per acre, exclusive of the standards which are left, and the crop, according to age and quality, will realize from £5 per acre downwards.

Great care is necessary to avoid damage to the stools when removing the fall, which is not usually done until just before the young buds are shooting out, and consequently at the time when injury is most easily brought about. The trampling of horses and passage of wheels are most injurious, while the browsing of cattle should be carefully guarded against. Good roads are always a great advantage in a coppice plantation, and to these as much of the produce as possible should be carried for loading, thus avoiding damage to the stools.

Although the growth of coppice wood has its disadvantages, particularly in park scenery, yet it is valuable in this way, that should the crop from local circumstances not be found remunerative, the plantation can at any time be converted into a standing wood by allowing the best and strongest shoots from the stools to form the permanent crop.

The approximate cost of forming a coppice wood per acre is as follows :—

	£	*s.*	*d.*
Trenching at 2*s.* per rod	16	0	0
5,000 trees at 35*s.* per 1,000 . . .	8	15	0
Pitting and planting	2	5	0
	£27	0	0

CHAPTER XVIII

THE FORMATION AND MANAGEMENT OF GAME COVERTS

When we consider that on not a few estates in this country the value of the plantations as game coverts is wellnigh of as much importance as that of the timber produced, it will be readily seen that the successful formation and management of these is a matter of no small moment to those entrusted with the work. That game-rearing and economic forestry can be advantageously carried on in the same woods is, however, a contention we by no means feel inclined to uphold, and from which, being apart from the subject matter of the chapter, we will for the present stand aloof.

Game coverts may be divided into two kinds, natural and artificial—natural when the woods are kept sufficiently thin to admit of the free growth of bramble, bracken or other native vegetation; and artificial when the planting of such shrubs as are suitable for underwood is resorted to.

Natural game coverts, which, by most sportsmen, are considered superior to those artificially formed, can only exist where the plantations are kept well and regularly thinned, so as to admit abundance of both light and air—the two principal requisites for the successful growth of natural underwood. Generally speaking, the formation of natural coverts has seldom to be helped, although occasionally it is found necessary to assist Nature by the sowing of such seeds as those of gorse, broom, etc., in the thinner and more open portions of the woodlands. This may, however, be considered an exception to the rule, as where the woods are kept sufficiently thin, spontaneous undergrowth is usually pretty abundant, and requires neither care nor management, beyond preventing its too free incursions along the margins of roads and shooting drives. Where, however, bare patches do occur, the sowing of seed may be relied upon as not only a speedy but most effectual method of increasing the cover. Where seeds are intended to be sown, the soil should be dug over, and all hard clods or lumps broken down, and the whole made smooth and fine with a rake. The seeds may be sown in spring, and afterwards covered over with hardwood branches as a preservative against the depredations of small birds and game.

The best natural game coverts are those composed of bramble, gorse, heath, hazel, holly, blackthorn, elder, blackberry, bracken or the stronger growing grasses, these being arranged according to merit, and each possessing some peculiar feature, specially recommending it for planting in certain soils, altitudes or situations.

In the formation of artificial game-coverts, when not only shelter and protection for game are required, but ornamental effect as well, the judicious grouping of the different shrubs should never be lost sight of, more especially when the coverts are within the park or policy grounds, and visible from drives and roads. Formality and stiffness are so often the characteristics of the present style of shrub planting, that in many cases our woodlands seem utterly destitute of that variety of outline and contrast of light and shade so essential to picturesque beauty. In planting evergreen shrubs for the two-fold purpose of covert and ornament, the best method is to plant each variety in separate groups or clumps. No hard and fast lines can be laid down as to the distribution or number of plants to be used in the clumps, which, to a great extent, must depend on the size and shape of the ground as well as taste of the operator. The clumps should, however, be placed at irregular distances apart, be irregular in size and outline, and with from forty or fifty to one hundred plants in each—bearing in mind that game of all kinds delight in small patches of shrubs with abundance of open space around each, but detest in a most marked manner continuous masses or jungles of underwood.

In selecting sites for the various groups, be careful to choose the most open positions, avoiding as much as possible planting immediately under the spread of trees; and, if practicable, so arrange that in viewing the wood from any point, the eye may not pass along a straight bare unplanted space, but become arrested by the various clumps in passing to the farther side.

Having arranged the positions of the various clumps, the pits should be opened of a size, and at a distance apart suitable for the plants intended to be used, taking care that they are sufficiently large to avoid cramping or bending of the roots, which in all cases should be spread out to their full extent. In making the pits, it is well to thoroughly loosen the soil in the bottom and sides with a pick, so as to give the tender rootlets a free course when starting into growth in spring. Should the soil be found of inferior quality, a few loads of leaf mould, road-scrapings or loam from an adjoining field will be found to work wonders in the way of giving the plants a start, and also in producing a strong, healthy growth. Drainage should also have been attended to previous to opening the pits, and all stagnant water or superfluous moisture removed by the formation of open ditches.

In giving a list of the best evergreen shrubs for covert purposes, I would call attention particularly to the merits of laurel, box, privet, laurustinus, rhododendron, holly and yew, as these have been very extensively used for underwood, and with the best possible results. As to which of the above shrubs should receive pre-eminence as an ornamental covert plant I cannot decide, each having some peculiar merit rendering it valuable in its own particular place. We will for the present, however, consider all alike in this respect, and briefly describe the value of each separately, beginning with the laurel.

The **Common** and **Colchic** laurels are amongst our best shrubs for underwood, and should be planted extensively; they are of free growth, bear cutting and pruning well, and thrive under the shade and drip of other trees. For covert planting the Colchic is perhaps preferable to the normal form, as it is of a more dense and procumbent habit, perfectly hardy, and less liable to injury from hares and rabbits. The common laurel requires frequent and heavy pruning to keep it in bounds, as, if allowed to ramble at will, it soon becomes bare near the ground, and useless either as game covert or ornament. Some years ago we layered a great number of this plant that had through neglect become useless for the purpose intended, many being from 12 ft. to over 20 ft. in height, with simply a tuft of foliage near the top. In layering, the stems were sawn half through near the ground, to assist in bending, and laid flat on their sides, a couple of stout pegs being driven alongside, the crooked heads of which served to keep the plants in their procumbent position. A spadeful of soil was then placed on the top of each peg to assist the layer in rooting. The result at the present time is everything that could be desired, each stem having thrown up quantities of young shoots, and thus formed a jungle of underwood, which year by year will increase in value.

In planting the laurel for covert avoid overcrowding, as, being of quick growth, the plants, even although placed at a considerable distance apart, soon unite and form a continuous undergrowth. No rule can be laid down as to the distance which should be allowed between individual plants, this depending entirely on their size, as well as on the quality of the soil in which they are to be planted. We not unfrequently plant double thick, either for immediate effect, or to produce covert at once, and when the plants begin to encroach on each other every alternate one is removed, thus giving the remaining plants ample room for developing side branches and thereby inducing a dwarf-spreading habit. Having a tendency, especially when confined, to increase more in height than width, the laurel, after a few years' growth, should have all the leading and straggling upper branches cut over, by which not only will the under shoots be increased but the plants will be prevented from running up into tall, branchless specimens.

The **Green Tree-Box** (*Buxus sempervirens*) forms a very pretty as well as desirable covert plant, and thrives well beneath the shade of deciduous trees. It is also of slow, dense growth, and well adapted for planting in various soils and situations, although preferring a light loam and a shady position. Another recommendation is its immunity from the attacks of game, hares and rabbits having such an aversion to this plant that even during the most severe weather I cannot remember having seen it badly injured. Few plants suffer more from overcrowding than the box, and for this reason it should be planted at wide distances apart, the plants soon getting top-heavy and falling over of their own accord. Where the plants are not of large size, and immediate effect or covert is required, they may be planted pretty close, and in a few years, when encroaching on each other, every alternate one may be removed. It is well adapted for transplanting, the almost solid mass of matted roots holding the ball of earth firmly together, thus rendering the plant one of our easiest as well as safest to remove.

The box would seem at one time to have been more abundant in our own land than it now is ; thus, Boxley in Kent, Boxwell in Gloucestershire, and Boxhill in Surrey, were named from the quantity of this plant which was formerly found in their neighbourhoods.

Privet, as a covert plant, has its advantages and disadvantages. On the one hand it is cheap, easily grown, and not at all fastidious about soil. When planted amongst trees, however, it generally assumes a loose, straggling habit, and as the shade increases it usually dies out altogether. Where the plantations are well thinned and regularly kept so, privet, if a little care and trouble be expended on its cultivation, will succeed and form capital underwood. In planting privet the greatest care is necessary to prevent its being overdone. Close planting is always productive of the most unsatisfactory results, not only as regards the health of the plants, but management of the woods as well. Instead of filling up the whole ground, as is not unfrequently done, plant in small clumps, and these at wide distances apart, as this will not only allow the privet to grow more healthy and compact but also admit of space for pruning and layering

—two necessaries for the successful cultivation of privet as underwood.

The layering of privet, which is a simple and inexpensive though effectual method of increase, is performed as follows :—Cut off all the branches, except those intended for layering, which are then laid flat on the ground equidistant around the main stem or root and kept fast by hooked pegs driven firmly down. A spadeful or two of soil should then be placed on the top of each peg, which will partly exclude air and hasten the formation of roots. The pegs may be made of any refuse branches—hard wood, such as ash or oak, being preferable—about 10 ins. in length, one end being hooked for holding the branches in position, and the other sharply pointed for ease in driving. As several forms of privet have crept into circulation of late, it is well to be sure that none but the true evergreen are used in the formation of game coverts. The oval-leaved privet, though a most desirable evergreen plant and well suited for ornamental hedges, is from its too luxuriant growth and upright form hardly to be commended for underwood ; at least, its merits in this respect are inferior to those of the common form.

Aucuba Japonica and the **Laurustinus** are two of our handsomest evergreen shrubs, but, unlike those already described, they will not succeed in the densest shade. In open places or along woodland drives they thrive well, and are excellent for variety and contrast. The laurustinus cannot, however, be considered as perfectly hardy in this country, for even in maritime situations where the air is to some extent ameliorated, it suffers severely from frost, and during severe winters it is even killed completely to the ground. It, however, springs very freely from the root, and in a few years quite regains its original size and luxuriance. From their bushy, well-furnished habit of growth both the above plants are excellent as game covert, more especially around the outskirts of woods and plantations. They should be allowed plenty of room for development of both root and branch, though they may, when necessary, be pruned with the greatest advantage.

Mahonia aquifolia, Berberis Darwinii and **B. Stenophylla** are frequently recommended as covert plants and for using in similar situations to those favoured by the laurel and box. Along the margins of plantations or in very open places they may and do succeed, but from practical experience of these plants we find them next to useless as underwood in shady positions. Where many thousands of covert plants are used annually, we have entirely discarded them from use except in the most open situations. These plants are highly ornamental, both in foliage and flower; produce berries which are much sought after by game, are quite hardy, and not at all fastidious about soil—qualities which specially recommend them for extensive use in positions at all suited for their growth.

The barberry, more especially when planted out in rich soil, and when at all confined, is apt to lose the compact, branchy nature so recognizable a feature of the plant when allowed ample room in the nursery border, and to assume a more upright habit of growth, which is anything but desirable in underwood generally. To check this and keep the plant in bounds, frequent light prunings will have to be resorted to, and this had best be effected during dull, damp weather, as the barberry is not a good subject for the pruning shears. Neither the barberry nor mahonia are adapted for planting in very high or exposed situations—at least where such has been tried the results have been anything but satisfactory, the plants soon presenting a miserable, half-starved appearance.

Both plants are readily propagated—the mahonia, when planted in loose soil and an open situation, soon covering a considerable space of ground, the running roots being especially active under such circumstances.

Rhododendron ponticum, although useful in an ornamental point of view, cannot be considered a first-class plant for game shelter. It has, however, several good qualities which recommend it for underwood, such as ease of culture, dwarf-spreading habit, and immunity from the attacks of game—indeed, in this latter respect, it is not equalled by any other plant, if we except one or two species of Daphne,

It is seldom resorted to by pheasants, the bottom being not only damp, but such a tangled mass of branches that it forms anything but pleasant quarters for game. For ornamental effect along the outskirts of plantations, the rhododendron is invaluable, and is by no means so fastidious about soil as is generally supposed, peat being not at all essential to its growth and successful cultivation. Few plants can be made to increase in like proportion with the rhododendron, and for this reason it should be planted in small patches; and when it is desirable to increase the cover, the outer branches may be pegged down or layered. This plant also bears pruning with impunity, so that old plants that have, through neglect, become lank and straggling, may without fear or risk be layered or pruned in with advantage.

The **Common Yew** and **Holly** cannot be too extensively used in the formation of game coverts, both being unrivalled for beauty and hardiness. They thrive in a great variety of soils, and beneath the densest shade of our woodland trees. In planting the yew it is well, however, to bear in mind that its branches are highly deleterious to farm stock that may browse upon them, and for this reason it should never be planted along the outskirts of a wood, or in any position to which such have access.

The **St. John's Wort,** as a low-spreading shrub, is unsurpassed, and thrives best in a light sandy or peaty soil. It is readily propagated by division of the roots; and when planted out in small patches a foot or two apart, the creeping stems soon cover a considerable surface of ground, and form a dense evergreen mass, covered in summer with bright golden flowers.

Gaultheria Shallon, another plant of creeping habit, is, notwithstanding its many good qualities, seldom planted to any extent in our woodlands; but this may, to some extent at least, be accounted for by the high price of the plants, and the small size of those purchaseable from our nurserymen. Like most other North American plants, the Gaultheria prefers a rather damp, peaty soil, and is one of the few shrubs found to thrive in pine plantations. The berries, which are borne in great abundance, are greedily devoured by pheasants, and in their native country are not unfrequently used as food.

The **Butcher's Broom** is a fine glaucous green shrub densely covered with sharp, prickly, leaves and invaluable for planting in shady places—indeed, in such positions it seems to be quite at home. There it flowers and fruits freely beneath half-standard rhododendrons where few other plants could exist, far less succeed. The twigs of this shrub were formerly used by butchers for sweeping their blocks; hence the English name.

Some of the above plants, notably the **St. John's Wort** and **Gaultheria,** may be considered as carpet plants, which, in contradistinction to general underwood, may be classed as evergreens, which, from their low, procumbent mode of growth, are scarcely in the true sense of the word suited for game coverts. To clearly define the difference would, however, be no easy matter, and, even were it possible to do so, would in the end be productive of but little good, as the habits of different plants vary so much that what is used in one place for carpeting purposes might in another and more favourable situation be equally valuable for game covert. A good example of this will be found in the St. John's Wort, which, when planted out and allowed to ramble at will amongst bramble, privet, etc., forms a capital covert; whereas, when used in open, airy situations—such as alongside shrubbery walks—it soon forms a dense evergreen carpet, of so compact a growth as to be almost impenetrable even to ground game.

In addition to the above-named plants, the following are well adapted for giving shelter to game:—Dogwood, Hazel, Elder, Arbutus, Cotoneaster of sorts, Juniper of sorts, *Pernettya mucronata*, *Rubus nutkanus*, *Taxus adpressa*, *Photinia serrulata*, *Kalmia latifolia*, *Garrya elliptica*, etc. These should be planted out in small groups—the more valuable kinds in the most conspicuous position, such as alongside or within view of woodland drives and shooting-roads.

Protection from Rabbits, etc.—It may seem somewhat absurd to speak of planting coverts, and then to protect them from the depredations of game; but that this is highly necessary for the first two years, at least, is well known to all planters. Few of the shrubs treated of in this paper are exempt from the attacks of hares and rabbits, more especially when in a young state and newly transferred from the nursery; and for this reason it is always found necessary to protect them in some way or other until fairly started into growth and beyond the reach of game. For this purpose wire netting is the cheapest and most effectual preservative with which I am acquainted. The netting should be about 4 ft. in height, not more than 1½ in. mesh, and inserted in the ground 4 in., to prevent rabbits from working underneath. It may be fixed to posts driven firmly into the ground at a distance of 5 ft. apart along the line of fence. This precaution against the depredations of game may not be necessary for all the clumps, but it is especially so for those of laurustinus, barberry, and laurel.

For the first two or three years after planting, the shrubs should be kept free of grass and weeds, as this will encourage the plants to start into growth more quickly and thrive much better than they can do if the ground is impoverished and light and air excluded by weeds.

CHAPTER XIX

HEDGES: THEIR FORMATION AND MANAGEMENT

STRICTLY speaking, hedges are divided into two kinds—useful and ornamental—the former being employed for keeping farm-stock in bounds, and the latter in the subdivision of private gardens and for lawn and park purposes generally. Where the fences are intended purely for protective purposes the thorn, beech, hornbeam or holly are the plants usually employed, while for ornamental garden subdivisions almost any shrub may be used, the choice of which will lie with the operator.

Amongst all the trees and shrubs that have been found suitable for the climate of Britain, none equals the common whitethorn, or Quick, for hedge-formation, where strength and shelter are points of first consideration. The beech and hornbeam certainly can thrive better on exposed and high-lying ground and where the soil is poor and thin, but neither forms so durable a protection against farm stock as the thorn.

When properly treated the thorn is a fast grower, and as a fence plant it is ornamental, smooth, stubborn, and long lived. It is also not at all subject to disease, and is very readily propagated. Few soils come amiss to the thorn—that is, if they are not overcharged with moisture, but it delights in a rich hazelly loam.

(1) **Plantation Hedges—Preparation of the Ground and Planting.**—Thorough preparation of the ground where live fences are formed should take precedence of all other operations—indeed, nothing can repay the planter more satisfactorily than the previous suiting of the land, in the

way of draining and trenching, to the plants intended to be inserted.

In all cases we have found it well to have the ground along the line of fence trenched to a depth of 2 ft., and about 3 ft. in width, and a quantity of manure incorporated at the same time. If this can be done some months before planting the thorns or other fence plants, so much the better, as it gives time for the loosely upturned soil to get mellowed

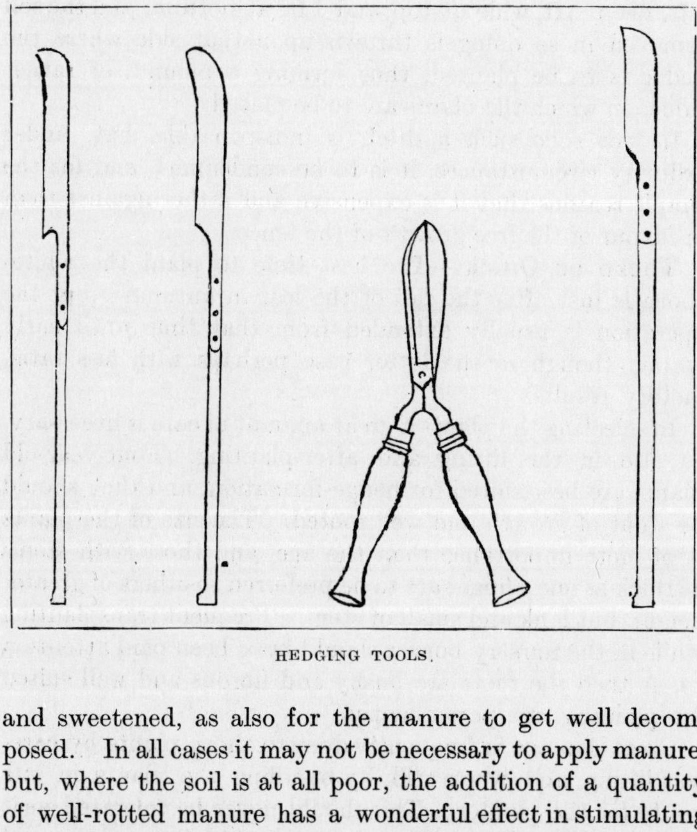

HEDGING TOOLS.

and sweetened, as also for the manure to get well decomposed. In all cases it may not be necessary to apply manure, but, where the soil is at all poor, the addition of a quantity of well-rotted manure has a wonderful effect in stimulating the quick into active and strong growth. The vigorous and rapid growth of a hedge, when the soil has been well worked and manured, is remarkable, and in all cases the labour and outlay expended on the fence is amply repaid.

It is well to plant rather above than below the general ground level, so that in trenching the soil a slight mound should be raised along the intended line of fence, which will not only materially assist in keeping the plants from excessive moisture, but aid in the cleaning and general management of the hedge. Where superfluous moisture is present in the soil the hedge-and-ditch system is to be recommended, which consists in digging out a ditch parallel with the line on which it is intended to place the fence. It should be 3 ft. deep, 5 ft. wide at top, and 1 ft. at bottom, and the soil removed in so doing is thrown upon that side where the hedge is to be planted, thus forming a mound, or rather ridge, on which the plants are to be placed.

In wet soils such a ditch is indispensable, but, under ordinary circumstances, it is to be condemned, and for the simple reasons that it is expensive and rather against than in favour of the free growth of the fence.

Thorn or Quick.—The best time to plant the whitethorn is just after the fall of the leaf in autumn ; but the operation is usually extended from that time until early spring, though in the latter case perhaps with less satisfactory results.

In selecting the plants a great amount of care is necessary, as also in the lifting and after-planting. Four-year-old plants are best suited for hedge-formation, and they should be stout of growth and well rooted. The size of the plants is of more importance than the age, and those with stems as thick as one's finger are to be preferred to others of greater height, but lank and small of stem. Frequent transplanting while in the nursery border should have been paid attention to, as then the roots are bushy and fibrous and well suited for planting out permanently.

Great damage is frequently done to thorn plants by careless lifting, and, worse still, by bundling the plants in lots ready for the planter. This should never be tolerated, as it is quite evident that when tied up in bundles and covered over with soil, the plants in the centre of each bundle get dust-dry and fall a prey to the searching winds of spring.

Plants should in all cases, where it is possible, be lifted

and replanted within the week, but, much better still—and this is readily effected where a home nursery is on the estate—on the same or the following day.

In planting, stretch a line along the centre of the prepared ground, and close to the line take out a perpendicular trench with the spade of sufficient size to allow of the roots of the plants being spread out to their full extent. From 6 in. to 8 in. will be found a convenient distance apart to place the thorns, and they should not be planted deeper than they stood whilst in the nursery border, which will readily be seen by the mark on the stems. A small quantity of fine soil should now be placed next the roots, and this firmly trodden, the remainder of the soil being added afterwards. Dibbling the plants is sometimes recommended, but, in our own opinion, it is a dangerous practice and to be avoided, the roots necessarily being thus confined to small space and placed in an unnatural position. Planting in single line is in most cases preferable to inserting in double line, as it is by the former method that the strongest and most durable fences have been formed. Some planters cut back the young thorns to within 2 in. of the ground, and the practice, although not readily reconciled with physiological principles, is to be recommended. It is unquestionable that headed-back thorns shoot out with greater vigour, and become thicker, than such as have been left untouched, but the fact that they are then greedily devoured by ground game has somewhat caused the practice to fall into disuse, at least where game is abundant and the cost of fencing cannot be entertained.

The **Beech**, as a hedge-plant, must not be despised, being a rapid grower on most soils, and soon forming a very valuable fence. In rich soils it retains a great proportion of its leaves during winter, and is, therefore, an excellent shelter-plant. It, however, lacks the rigidity of the thorn, and for that reason is not very suitable for planting where farm stock have access. It may be planted in a manner similar to that recommended for the thorn, only the individual plants should stand farther apart.

The **Hornbeam** makes a good live fence, and will grow

readily in any fairly good soil and not too exposed ground. It may be treated similarly to the beech.

Privet—both the common and oval-leaved—have been largely used, either alone or with other plants, in the formation of hedges, for which they are peculiarly suitable. They, however, want stiffness, so as to be able to cope with farm stock, and for this reason are principally used in ornamental garden sub-divisions.

Gorse or Furze.—Strikingly beautiful as well as useful hedges may be formed of gorse. It is well adapted for planting on light dry or sandy soils, or on the top of a dyke or sunk fence.

Seed sowing is to be recommended in the formation of gorse fences, and after preparing and well working the soil, 1 lb. of seed to every 100 lineal yards will be found sufficient for sowing down. It should be remembered that in order to keep the fence full and bushy, pruning should take place immediately after flowering and before seeds are produced.

Cutting over the hedge at ground level every third year will be very beneficial to this fence.

(2) **Ornamental Hedges—Holly.**—This makes an excellent ornamental fence, and it is occasionally though rarely used for plantation purposes. The ground should be thoroughly prepared, and, if necessary, enriched by a dressing of strong loam, and the plants inserted in May. The holly can be planted when of almost any height, if previous transplanting was attended to.

Yew.—For purposes similar to the latter, the yew is generally in use. It may be planted at any time, but should be kept well watered until it has become established.

Laurustinus.—As a flowering hedge for garden or lawn purposes few shrubs are of greater value than the laurustinus, particularly in maritime districts. In severe winters it suffers considerably, though generally fresh growths are sent up from the rootstock. Pruning should be carefully done, so that the flowering shoots are not cut away.

Box.—Very neat and serviceable garden fences are made of this shrub. It grows freely and stands pruning well.

Rosa rugosa and **R. rubiginosa** (Sweet Briar).—Both

these species of rose have come greatly into favour for garden subdivisions, for which they are peculiarly suitable. They make charming hedges, are of easy growth, and stand pruning with impunity. Of course, where they are wanted to flower great care in pruning is necessary.

Laurel.—Both the common and Colchic laurels make fairly good hedges, but they are apt to get gappy by portions dying out. This can, however, easily be remedied by filling up with others instead. They bear trimming well.

Aucuba japonica forms a useful and ornamental hedge, stands pruning well, and lasts for a long time.

Berberis Darwinii and **B. stenophylla** are both highly ornamental hedge shrubs, and when not pruned too severely flower with great freedom. Shortening the long shoots with a pocket knife is best.

Cleaning and Pruning.—An annual cleaning of the ground alongside hedges must never be neglected, as weeds rob the soil of its nourishment, choke the young plants, and to a great extent prevent the free access of rain to the roots of the hedge plants. The common hoe is, for this purpose, to be recommended, and any weeds that cannot be got at around the stems must be removed by the hand. Ivy, elder and honeysuckle should all be treated as weeds, for they are highly injurious to hedges, be these young or old.

Very little, if any, pruning or switching should be done till the third year after the fence has been formed, and then only the longer twigs cut back, so as to get by degrees a general uniformity of shape. The switching-knife is alone to be recommended for pruning fences, shears never making a clean cut, and pressing and loosening the bark at the point where amputation took place. A well-sharpened switching-knife in the hands of a dexterous hedger turns out beautiful and commendable work.

Unless it be an annual cleaning and trimming, a well-formed hedge should require but little attention for many years. Should it, however, when old, begin to show signs of distress and become gappy, a top-dressing of rich farm-yard manure will go a great way towards throwing fresh energy into the plants. This should be applied in winter, and lightly forked in the following spring, and before growth has commenced. Should gaps occur by reason of deaths in the old plants these should be removed, and others of young growth substituted, the soil at the same time being dug out and other fresh from a field or roadside used instead. Great care should be exercised that the roots of living plants are not injured whilst removing the dead and substituting the live specimens. Specially-prepared plants and such as are unusually stout and bushy should alone be used in hedge-repairing.

CHAPTER XX

SHRUBS FOR SHADY SITUATIONS

THE list of procumbent or carpeting shrubs that have been found to succeed when planted beneath the shade and drip of forest trees is by no means a long one, and as the clothing of such bare places is often a point of the greatest importance, particularly in park or ornamental grounds, the following notes may prove useful. By careful manipulation much may be done to carpet such places with suitable shrubs, but the task in some instances is by no means an easy one, and must be set about in a common-sense and practical way. Evidently deciduous trees have an advantage over evergreen kinds in that with the extra light and greater amount of surface dampness undergrowth succeeds better beneath their shade.

The **common Ivy** is probably the best evergreen carpet for shade planting with which we are acquainted. It runs about and roots freely, soon covering a large space of ground with its neat, deep-green foliage. Propagation is brought about either by means of cuttings or suckers, and is simple and inexpensive.

The **Periwinkles** (*Vinca major* and *V. minor*) are well adapted for planting beneath our larger trees, where, unless the shade is very dense, they succeed admirably, soon forming large breadths of evergreen carpet and producing their deep blue flowers in abundance. They are readily increased by layering or subdivision, and when once established soon spread about unheeded.

The **St. John's Wort** (*Hypericum calycinum*) can confidently be recommended for planting as a ground covering beneath our larger trees. It increases readily, and if occasionally cut over, shoots out all the more freely and thickly.

For the showy yellow flowers it is also a desirable shrubby plant.

The **Mezereon** and **Spurge Laurel** (*Daphne Mezereum* and *D. Laureola*) are excellent, medium-sized shrubs for planting in shady positions, where they not only succeed well, but flower freely. They are both increased by layering.

Euonymus Radicans Variegata is a useful, procumbent shrub for planting in the shade, and succeeds well in smoky localities.

The **Butcher's Broom** (*Ruscus aculeatus*) grows with great freedom beneath the densest shade of our forest trees, and being an evergreen is to be recommended for such situations.

Gaultheria Shallon and **G. Procumbens** may also be recommended for planting where the shade is not too dense ; they both flower and fruit freely, and are of neat procumbent growth.

Amongst taller growing subjects for planting in the shade, mention may be made of the holly and yew, both of which thrive beneath the shade and drip of forest trees and where they often assume a dwarf, procumbent habit of growth. Two at least of the recent shrubs introduced from China have just claims to be included in the list of suitable species for planting in the shade. These are *Sarcoccos humile* and *S. ruscifolia*. They are of particularly neat growth, with persistent leaves and bear yellowish white flowers.

The **Common Ling** or **Heather, Blackberry,** and **Andromeda Catesbœi** all succeed well in the shade of trees, particularly if the soil be inclined to peat.

It frequently happens that the soil beneath large trees is thoroughly exhausted, and that the small, fibrous rootlets are so abundant as to render planting almost impossible. Under such conditions it is advisable to first gently loosen the soil, without disturbing the larger roots, and add a top-dressing of, say, three or four inches of good friable loam. This, thoroughly incorporated with the existing soil, will give the young plants a start and allow of their becoming strong and established before the encroaching roots rob the ground of its nourishment.

CHAPTER XXI

INSECT ENEMIES OF TREES

WHETHER viewed from a commercial or sylvicultural point of view, the widespread damage caused to timber from insect attacks can scarcely be overrated. The depredations in various parts of our own country, particularly in young plantations, are bad enough, but when compared with those of Europe and America they appear insignificant. In France and Germany whole woods have been wiped out by insect pests, while the Government of Bavaria were mulcted in something like £100,000 by the destruction of its spruce forests. The United States fares no better, for we find that over a period of ten years the amount of timber killed and reduced in value was calculated at fully £10,000,000. The coffee plantations of Ceylon suffered much from the attacks of a fungus, and we could go on multiplying cases. In our own country the ravages of the pine beetle and of the larch disease have caused incalculable damage; indeed, in the latter case there is hardly a plantation of larch where the presence of the fell disease cannot be distinctly traced, while the pine beetle has ruined whole plantations both in England and Scotland. Though the adult bark and wood-boring beetles do a great amount of damage, yet that inflicted by the caterpillar or grub from the egg is greater still, and in the case of fungi we have a typical example of their destructive properties in the case of the well-known larch disease.

The **Pine Beetle** (*Myelophilus* (*Hylurgus*) *piniperda*) is a dreaded enemy to not a few species of Pinus, but particularly *P. sylvestris*, *P. laricio*, *P. austriaca* and *P. Strobus*. The injury done by this beetle consists in its destruction of the leading shoots of the tree it attacks. It enters by boring a hole into the side of the shoot until it reaches the pith, after which its course is directed upwards, and an exit made at the terminal bud. This tunnelling of the shoot so weakens it that frequently during stormy weather it is broken across at the point where the beetle entered. Not only are unhealthy trees attacked by the pine beetle, but young and robust-growing specimens frequently fall a prey to its insidious depredations.

June, July and August are the months when it is most commonly found.

The only remedy is to collect and burn the affected shoots—work that requires to be done with the utmost care to prevent the escape of the wary insect. Burning all brushwood in plantations is a great preventive.

The **Pine Weevil** (*Curculio* (*Hylobius*) *Abietis*) is another destructive insect, which differs from the former in waging its attacks against the buds of the leaders and branches, as also by eating patches of the bark here and there on the stems and branches. The various species of Abies suffer most, but the pines are occasionally attacked as well. It is always most destructive in young plantations growing on the margins of old woods, and equally bad amongst trees that have been planted on the site of a former pine plantation.

The beetle is about half an inch long, and nearly black. One remedy, probably the best, is to place fresh pieces of pine bark on the ground, beneath the infested trees. By shaking the trees and examining the traps the following morning, many may be destroyed.

Bostrichus typographus is another pest of our woodlands, and may frequently be seen, like fine white wool, spreading over the stem and branches of the silver and other firs. It spreads with terrible rapidity, first appearing in small patches here and there on the bole, and particularly on the under sides of the branches. The tree infested soon becomes unhealthy, and frequently dies off prematurely. Trees growing in low-lying, heavy ground would seem to fall a first prey to this insect.

Bostrichus laricis is nearly allied to the former, but its devastations, which are, however, not very deadly, are principally confined to the larch. It is usually known as the "larch blight."

The **Pine Shoot Moths** (*Retinia buoliana* and *R. turionella*) would seem to be more numerous in this country than is generally supposed. Quite lately I visited a large plantation of young Scotch fir, the terminal buds of which were greatly injured by the caterpillars of this elaborately-coloured moth. The moth lays its eggs at the base of the buds, and into these the caterpillars enter by hollowing out the centre, thus destroying their vitality and causing them to take on a withered appearance and to feel soft and empty to the touch. Trees infested by this insect resemble greatly in their stunted shoots and exudation of resin such as have become a prey to the Pine beetle (*Myelophilus piniperda*), only in the latter case it is the fresh young shoot and not the bud that is attacked. The *Retinia* would seem, from all my notes and observations, to be most abundant in what might be termed neglected fir plantations, that is, where the trees have suffered from overcrowding, or from unfavourable conditions as to soil, etc., and particularly when the wood is composed entirely of one species. There is no method of dealing with large infested areas, for the attacked trees have repeatedly been cut over and removed without any seeming diminution in the numbers of the insect. One experiment with a small infested corner has been rewarded with good results, viz. the lighting of a fire to windward, and causing the smoke of coal tar to pass over the infected area. This might be worth trying in the case of fruit trees infested by particular insects.

The **Larch Miner** (*Coleophora laricella*).—Few, other than those specially interested in tree diseases, have the remotest idea that the yellow, withered appearance of many of our English larch plantations is due to the larvæ of the above tiny moth. It usually attacks young trees, say, from five to twenty years old, and although it may not kill them out, yet the repeated onslaughts year after year tend to keep the trees in an unhealthy condition, and so render them liable to other and more deadly diseases.

Unfortunately the attacks of the larch miner are by no means confined, as is usually supposed, to trees growing under unfavourable conditions, for I have this season noticed in an unusually healthy, fast-growing plantation in Sussex that almost every tree was more or less affected. Certainly in another large extent of larch in Gloucestershire which I examined lately, where nine-tenths of the trees were being ruined by the *Peziza*, the larch miner was very abundant ; but, I think, that young trees, whatever may be their state of health, suffer alike, although where hard-wooded trees form a portion of the crop the larch certainly suffers less than when grown in pure woods. The moth lays its eggs at the end of June on the needles of the larch ; the caterpillar mining into and feeding upon the interior of the needle causes it to turn faded and yellow. It lives in the tube thus formed during the winter, changing to a pupa, and ultimately to a moth. It is a most difficult matter in the case of this insect, as, indeed, of all others that are fairly abundant, to suggest a remedy, and I have looked over and examined larch plantations that are differently situated in many respects to find out under what condition the attacks are most persistent, but with little or no success—healthy and unhealthy, native or Tyrolese, faring alike when grown as a pure crop.

Where the larches are intermixed with hard-wooded trees —sycamore, oak and beech—the attacks are certainly less frequent, as I have noticed in a number of cases. Trees growing at high altitudes do not seem to suffer less than those only a few feet above sea-level, and this point I have paid particular attention to.

Whether the wounds caused by this insect will serve as a nidus for the spores of *Peziza Willkommii* has yet to be determined, but special importance should be attached to all larch-feeding insects, and their depredations minimized to as great an extent as possible.

The Pine Sawfly (*Lophyrus Pini*).—Fortunately, this insect is not abundant in the British Isles, though on the Continent the damage it does in the pine forests is by no means inconsiderable. The insect may readily be recognized by its wide, flattish body, and usually dark appearance. Having attained to full size in the trees, they form cocoons among the foliage or on the stems, and remain in this condition until the following spring, when, in April or May, the perfect insects make their appearance. The male is considerably smaller than the female, while the full-grown caterpillar, which is of a greenish-yellow colour, with a row of black spots on either side, is about an inch long. The remedial measures are not at all easy, especially when a large number of trees are attacked, but single specimens may be entirely cleared by shaking the caterpillars into a sheet placed beneath the tree.

The **Larch Aphis** (*Adelgis laricis*) and **Giant Sirex**

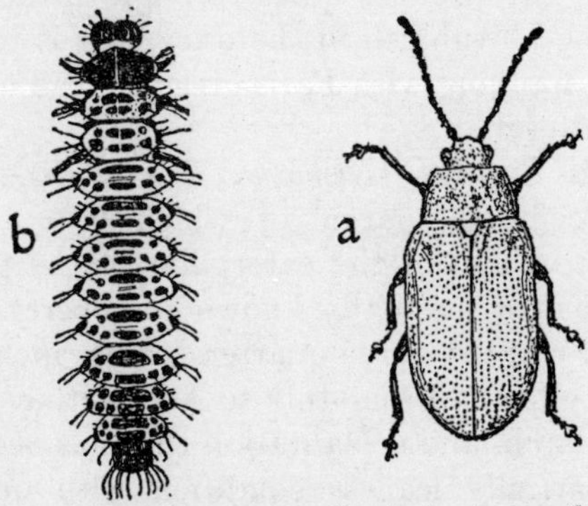

(*a*) WILLOW BEETLE (*Phyllodecta vulgatissima*), AND (*b*) ITS LARVA
Magnified 5 times

(*Sirex gigas*) are both, more or less, harmful to the larch. The latter is a formidable and splendid insect, which is, however, not very abundant in this country.

Generally felled trees, or such as are somewhat sickly, are chosen by the female in which to lay her eggs. These are deposited beneath the bark by means of the powerful ovipositor, and in course of time the whitish cylindrical maggots make an appearance, and with their strong jaws form large borings in the affected tree.

Cutting down and burning infested trees is the only practical remedy.

The **Willow Beetle** (*Phyllodecta vulgatissima*) causes

considerable damage to osier plantations, and would appear to be greatly on the increase of late years. Much damage has been done to osier holts in various parts of the country, and in northern Ireland the ravages of this beetle were particularly noticeable during the past five years. The insect, which is metallic green or blue in colour, passes the winter in the adult state, at which time it may be found amongst refuse of the osier beds, such as the heaps of bark, and also at the base of old stools and beneath stones or other shelter. The larvæ have a tough yellowish cuticle with conspicuous brown bristles, the head and prothorax being black and hard. The eggs are laid on the undersides of the leaves in spring, and when the larvæ are hatched they feed on the leaves, eating holes quite through to the upper surface. Burning all rubbish in the osier beds is to be recommended, and spraying with Paris green or lead arsenate has been found useful.

The **Larch Sawfly** (*Nematus Erichsonii*).—This is a species of sawfly the larvæ of which bear considerable resemblance to those of the caterpillar of the pine sawfly, and also to that of the better known gooseberry caterpillar. The larvæ are about three-quarters of an inch long, and possess twenty feet. From July to August they feed on the leaves of the larch, and a plantation that has been attacked presents a partially leafless condition with quantities of the brown cylindrical cocoon cases lying amongst the grass beneath the trees. There have been several notable instances in which larch plantations have suffered severely from the attacks of this insect, and in northern England, particularly Cumberland, whole areas of plantation have been attacked. Burning all brushwood and grass beneath the trees in infected plantations is probably the best means of lessening the numbers of this dread insect in our larch plantations.

The **Spruce Gall Aphis** (*Chermes abietis*).—This is a common insect, and one that renders many fine young spruce trees very unsightly by reason of the cone-like excrescences that are formed by the action of the insect on the shoots of the infested specimen. The formation of this excrescence is brought about by the female aphis piercing with her beak, or sucker, one of the buds, and drawing off the sap, the consequence being an unusual growth at that part.

When the young larvæ appear, they also, by piercing the

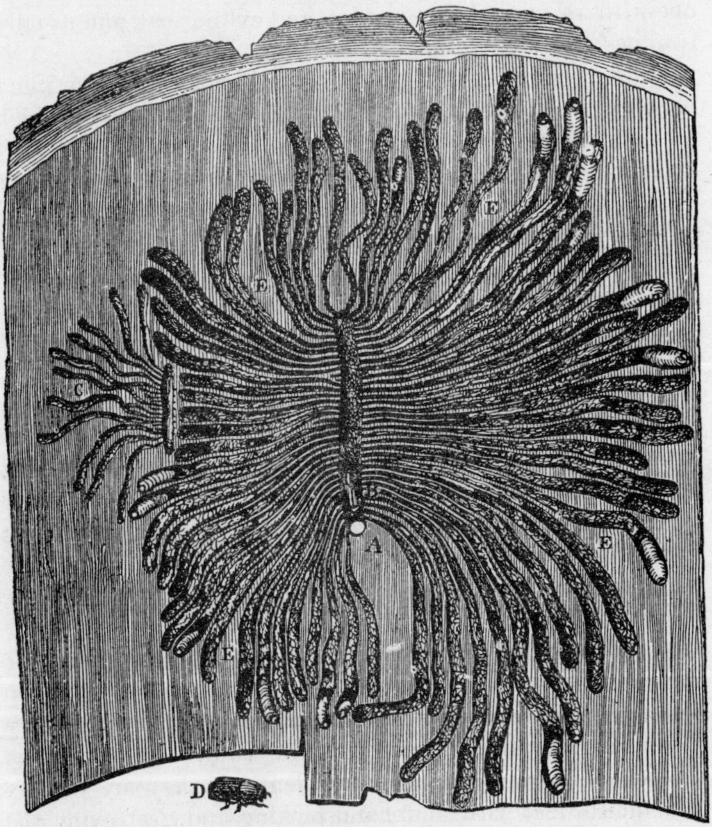

ELM TREE DESTROYER

A. Entrance of Parent Tube.
B. Insect in Tube.
C. An imperfect Tube arrested from want of nourishment.
D. Scolytus Destructor. Natural size.
E. Lateral Tubes with Larvæ, some changing to the Pupa state.

gall, extract the juices, and the gall enlarging soon causes the larvæ to become embedded at the bases of the leaves, which, by this time, have become curiously malformed. The insects are scarcely one-tenth of an inch long.

The only remedy is to collect the cone-like excrescences and have them destroyed, except in the case of badly-infested trees, which should be cut down and burned.

The **Elm Tree Destroyer** (*Scolytus destructor*) is about one-fifth of an inch long, stout and cylindrical, and usually confines its depredations to the elm.

In the beginning of June this beetle bores into the inner bark, where it forms galleries, along the margins of which are laid the eggs. Cutting down and burning badly-attacked trees is the best remedy, but promoting exuberant health of the infested specimens by means of enriching the soil has been attended with promising results.

The **Goat Moth** (*Cossus ligniperda*) is most frequently found on the willow, oak, lime and other trees. Being not only one of our largest native moths, but also one of the most destructive, its ravages are much dreaded, the holes or tunnels made by the moth being of large size—large enough to admit the little finger. Filling up the tunnels with a mixture of soot, lime and cow manure is an excellent remedy.

The **Lackey Moth** (*Clisiocampa neustria*), so called from the gay colours of the caterpillàr, is another destructive woodland pest, eating wholesale the leaves of the oak, elm, beech, poplar and most fruit trees.

In April and May the caterpillars are hatched, when the leaves are just unfolding. They form a nest or web of silken hairs, generally amongst the smaller branches, in which they live during the day, sallying forth in the evening to feed on the tender foliage. Being very plentiful, they are usually difficult to deal with, but hand-picking and destroying the cocoons are the only practicable methods of meeting the evil.

The **Winter Moth** (*Cheimatobia brumata*) and the **Lime Looper Moth** (*Hybernia defolaria*).—The caterpillars of both these moths are very destructive to the leaves of elms, limes and willows, but particularly to the buds of the apple tree.

When full grown they descend to the ground, where they cover themselves and become chrysalides, from which the moths appear from October to December.

Being almost wingless, it is by no means difficult to prevent their ascending the trees by painting a band of any sticky substance around the stems of the trees that it is expected they might attempt to crawl up.

A

A. Larva of the Cossus Ligniperda, three years old, ready to change into the Chrysalis state.

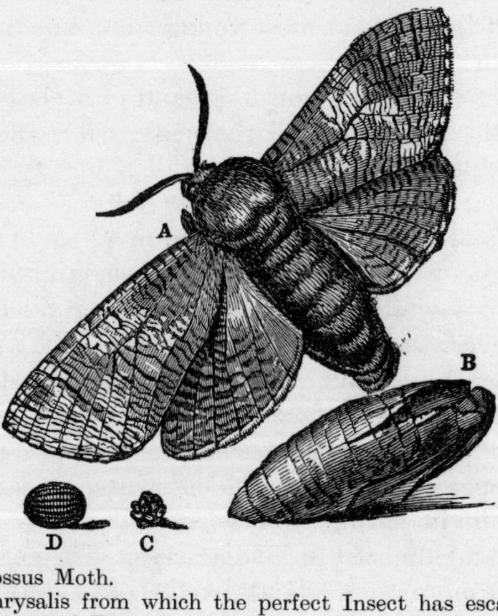

A. Cossus Moth.
B. Chrysalis from which the perfect Insect has escaped.
C. Cluster of Eggs.
D. Magnified Ovum.

THE GOAT MOTH

The **Red Spider** (*Tetranychus*).—In hot and dry summers trees suffer much from this member of the mite family. Limes and poplars, as also many other trees, are greatly injured, the foliage turning to a russety brown colour, and falling off long before the usual time. There are several remedies, such as fumigating and spraying with a solution of soft soap, but none of these are applicable to a plantation of trees, or even a single specimen of large size.

The **Thorn Fly** (*Aphis Cratægi*) attacks whole hedges or brakes of Quick, especially those in the nursery border; as a rule, the younger and more healthy plants first fall a prey to its depredations.

Sponging with tobacco water, or almost any of the prescribed solutions will rapidly exterminate the fly; but such work is laborious when a large brake or long hedge of the thorn has to be gone over.

The **Cockchafer** (*Melolontha vulgaris*) is usually pretty abundant, and does most damage by eating the leaves of the sycamore, beech, oak, cherry and many other trees. It will also eat the roots of most young trees, but those of pine in particular.

The insect is about 1¼ in. long, and of a chestnut-brown colour on the upper part of the body, while the head and some other parts of the body are of a bronzy green, and thickly covered with yellowish-white hairs.

In April and May the eggs are laid in a hole in the ground about 5 in. deep, and the grubs are hatched in July. They are of a dirty-white colour and much wrinkled. In this state, however, they do but little harm; but, after having changed their skins and remained in a torpid state during winter, come to the surface in spring and eat the roots of almost any plant that comes in their way. They again burrow deeper at the approach of winter, coming to the surface again in spring, and, when full grown, are about 1½ in. long, and almost ½ in. in diameter. The perfect insects do not live more than about twelve days, and are easily known by their heavy, awkward flight towards the evening.

The **Laburnum Moth** (*Cemiostoma laburnella*) is fairly abundant—in England, at least; and, in some instances, every leaf of a tree has been eaten almost wholesale by the caterpillars of this pretty moth. The insect is about one-eighth of an inch in length, and three-tenths of an inch across the fully expanded wing. It is of a silvery-white colour.

The greenish-grey caterpillars are about ¼ in. long.

By burning the attacked leaves great numbers of the caterpillars may be destroyed, while, by shaking the trees in May and August, the moths will fly out, and may be caught in a butterfly net.

Wood Leopard Moth (*Zeuzera æsculi*).—The caterpillar of this beautiful moth is very destructive to the beech, ash, birch, elm, walnut, privet, etc., which it bores into, eating and living on the wood. Usually young trees or the branches of old specimens are attacked, and the tunnelling

WOOD LEOPARD MOTH

(*Zeuzera æsculi*)

(Very destructive to trees, particularly around London)

is confined in the former either to the pithy centre or the soft wood near the bark. The moths appear about mid-July, and the female, by piercing the bark with her powerful ovipositors, deposits her eggs, one in each hole. Three years are required for the caterpillar to arrive at maturity when it is nearly 2 in. long. Both in form and colour the leopard moth is particularly elegant, the head and thorax being covered with a thick white pile, the body with a black down, fringed with white at each joint. The wings are white with yellowish-brown veins, a row of rounded bluish spots running between every two.

By stuffing a piece of tow in gas tar, or placing cyanide of potassium in the hole and closing the aperture, the caterpillar may be overpowered and destroyed.

A bent wire has often been successfully used in dislodging the caterpillar.

The **Holly Fly** (*Phytomyza aquifolia*).—The foliage of the holly is frequently very much disfigured by the grubs of the holly fly, which burrow beneath the upper skin of the leaves, feeding on the internal substance. This imparts a blistered and discoloured appearance, which, in the case of ornamental varieties, is anything but desirable. Probably no great damage to the infested trees is brought about, but the wholesale destruction of the leaves, as is often the case, cannot but weaken the plant.

In May and June the flies make their appearance, and lay their eggs beneath the upper skin of the leaf, from which the grubs, about one-fifth of an inch long, are hatched. These work their way beneath the skin of the leaves, forming small tracks of a more or less circular shape, thus causing the large and unsightly blisters. They quit the leaves about March, by making small holes in the skin of the leaf, and afterwards become chrysalides. The fly is small and inconspicuous.

Picking off and destroying affected leaves, or crushing the grub by pinching the blisters are the only ways of lessening the attacks.

The **Oak Leaf Roller Moth** (*Tortrix viridana*).—The widespread destruction caused to oak-woods in almost every part of the country by the caterpillars of this little moth would seem to be on the increase from year to year. But it is not the oak alone that suffers, for numbers of the hornbeam and beech are in an equally pitiable condition. Having closely watched this insect for several years, mainly with the view of striving to keep it in check or devise some means of destruction, I have come to the conclusion that a few individual trees may, at considerable expense, be rid of the pest, but in the case of whole woodlands artificial treatment is quite out of the question. Many observers are under the impression that the caterpillar is most abundant where the trees grow closest together, and when we consider that both wind and rain destroy numbers of these, the fact of isolated trees, which, consequently, are most exposed to storms, being comparatively free from attack is not to be wondered at. So far as we know at present, the only way to diminish in any appreciable degree the numbers of this insect is by encouraging as much as possible its enemies.

The rook, jackdaw, starling, thrush and sparrow help immensely in destroying the caterpillars, and the occupants of a rookery will frequently in a few hours clear the pest from the trees over a considerable area of woodland. The same has been noted with regard to the starling, and I have frequently seen the trees over an infested area almost black with this particular bird when in the act of feeding on the caterpillars.

In support of this recommendation it may be stated that insects are far less numerous in the forests of St. Germain, Senart and Fontainebleau than in the Bois de Boulogne, where, of course, small birds are scarcer.

Sometimes with the oak leaf roller moth its excessive number proves the means of its extinction, the foliage being devoured before the caterpillars are fully fed ; while, as is usual at the season of attack, parasitic flies and ichneumons destroy them wholesale, and a box of caterpillars sent to me the other day revealed the fact that each one had succumbed to the attacks of one of these enemies.

The life history of this moth is full of interest, and the curious manner in which it rolls up the leaves is well worthy of study.

The Felted Beech Coccus (*Cryptococcus Fagi*).—Judging from the numerous specimens of the beech coccus that are being forwarded to me for identification, and the inquiries as to how this insect pest may be dealt with, its presence in almost every part of the country is indicated. It was only in 1862 that Dr. Balfour reported the presence of this coccus in Scotland, and I remember well how the beautiful beech hedges on the Penicuik Estate, Mid-Lothian, were ravaged by the insect in 1875. In Germany, however, the beech coccus was noticed as early as 1849. It is probable that the beech coccus also extends to the Weymouth pine, as at Keston, Lord Derby's estate in Kent, several of these trees were badly affected with an insect that appeared to me identical with this pest, and other instances of the Weymouth pine being similarly attacked have been recorded. Since the appearance of the insect in Scotland it has spread southwards rapidly, but it is only of late years that its ravages have been felt severely, and that owners of woodlands have been driven to do everything in their power to combat its injurious effects. I think I have nowhere seen the insect so abundant as on the Burnham beeches, which I visited in company with several members of the Royal Horticultural Society at the request of the Lord Mayor and the Corporation of the City of London. Here, speaking broadly, all the trees are affected in a greater or less degree, independent altogether of the age or health of the trees.

Amongst the beautiful beech woods of Hertfordshire, where the timber produced commands a higher price than that from any other station in Britain excepting the Chiltern Hills, the trees are suffering severely. After a careful inspection of many beech woods, I am at a loss to account for the spread of this insect. Usually, in the case of injurious forest insects, the presence of dead and dying wood and the general health of the plantations have much to do with the attack, but this would not appear to be the cause in the case of the beech coccus. Even old and diseased trees do not appear to suffer more than the young and healthy specimens, and at Burnham some of the youngest and fastest growing specimens were by far the worst affected, appearing in parts as if coated with drifted snow. Neither would soil appear to have anything to do with the spread of the insect, as on chalky, sandy and loamy soils, as well as on shale rock, the trees are all more or less affected. I cannot say that unhealthy trees are more liable to be attacked than vigorous ones, for I have observed diseased trees to be sometimes quite free from the coccus, whilst healthy specimens in the immediate neighbourhood were badly infested. This was particularly noticeable on some park trees on two estates that I visited lately; in each instance the majestic boles rose to a height of fully 90 ft. and contained about 100 cubic ft. of wood. At one time I had an idea that beech trees when grown too thickly were most liable to be attacked, but more extensive observations have disproved the hypothesis. By some it has been suggested that the drainage of ground on which the beech is growing will facilitate the spread of the insect by bringing about an unhealthy state of the trees. To some extent this has been noticed at Burnham, and particularly in Gloucestershire, where a lake had been formed contiguous to a beech plantation. In another instance that came under my notice the burrowing of rabbits to an unusual extent was blamed for the appearance of the insect and the gradual death of the trees, the soil in this case being sand and gravel. It is a strange fact, too, that a badly infested tree may be standing amongst others that are perfectly free from the attack. Under a magnifying glass the insect appears of a yellowish colour, irregularly globular in shape, and almost transparent. It is thickly covered over by a secretion from the body, which looks like fine waxen-white threads or cotton wool, which effectually screens the insect from observation. With its proboscis it sucks up the juices, which are readily reached through the thin, smooth bark. When badly attacked, the foliage becomes meagre and has a burnt appearance, then the tips of the branches, and finally whole branches die off. Afterwards the bark becomes dry and cracks, usually longitudinally, falling off in flakes with the death of the tree. The timber of trees that have been killed by the coccus is dry, short-grained, and by no means comparable with that of healthy specimens. From whatever cause, the presence of the insect in our plantations is much to be deplored, and already many fine old trees in different parts of the country have succumbed to the attack. It is a curious fact that certain trees fall victims to the attack of this insect very rapidly, often in the short space of two years, while others of equal age and vigour, and, as far as can be seen, similarly situated, live for many years. Fully grown trees are, in my opinion, more liable to be killed quickly than younger specimens.

As regards remedies, these can only conveniently be applied to single specimens and are scarcely practicable on a large scale. That success has, however, attended the application of certain remedies I am quite convinced, for several valuable ornamental trees on a lawn in Buckinghamshire that were badly affected by the pest are now, twelve months after being treated, quite free from the insect, and do not appear to have suffered in consequence. In this case the remedy was simple, and consisted merely in scrubbing the tree stems where affected with a stiff brush dipped in an emulsion of soft soap, this being rubbed well into all crevices of the bark. Only one application was given, but as the trees were nearly 100 ft. high with many large limbs, which also showed the presence of the coccus, the work took a considerable time. However, the results well repaid the expense of labour. Another excellent remedy is to mix together equal portions of paraffin and soft soap, and when required for use add twenty times their bulk of hot water, stirring all well together. This may be applied either with a scrubbing brush or syringed on to the affected parts. Other measures of a more drastic nature have been found to be beneficial, but those given are probably the simplest and best. Three or four handfuls of lime to a bucket of water is an excellent application, the only drawback being the unpleasant colour imparted to the trees. This is a simple remedy, and may be applied with a painter's or scrubbing brush. It has saved many trees on an estate near London.

Wireworms.—These occasionally do a great deal of damage in beds of seedling trees, particularly conifers, and in some instances they attack and destroy the seeds before germination. In the case of young conifers they are gnawed completely through just above or at the ground level, the beds in many instances being strewn with the cut-over plants. *Abies nobilis* and *A. Nordmanniana* suffer to a great extent, and I have frequently been at my wit's end to put a stop to the repeated depredations. In the case of a newly-formed nursery or freshly-made-up seed beds the attack of the wireworm is always most pronounced. In the case of fresh nursery ground, paring off and burning a couple of inches of the top soil in the autumn has been attended with excellent results, as has also dressing the ground with gas lime. When seedlings are attacked, hand picking, with the use of sliced carrots, mangold or potatoes, varied with pieces of oilcake as a bait, are to be recommended. Dress-

WITCH'S BROOM ON THE WILLOW (SUMMER APPEARANCE)

ing the seeds with red lead is very advantageous to prevent birds and mice from attacking freshly-sown quantities. Injury from wireworm is not likely to cause any serious consequences after the first year's growth of the plants, especially if the ground has been kept clean and free from weeds during the egg-laying season in June.

The above are only a few of the many insects that injure our forest trees ; but those treated of are the most familiar to the forester, and those whose attacks he has most frequently to wage war against. Willow trees all over London are suffering severely from attacks of a mite insect. It is known as the "Witch's Broom" on the Willow (*Eriophyes triradiatus*).

Remedial measures are frequently of little avail, more particularly when a whole wood or plantation is attacked;

WITCH'S BROOM ON THE WILLOW (WINTER ASPECT)

but with single specimens the numbers may readily be lessened by the methods suggested.

Keeping the woodlands free of dead and dying trees and branches is a preventive of insect attacks that should never be neglected.

As showing the injury that can be inflicted on our forest

trees by some of these insect pests, the following cases may be mentioned :—

During fifteen years (between 1853–68) the spruce in East Prussia, Poland and Russia was killed over an area of 7,000 square German miles ; while in the Bavarian forests, the loss to the Revenue in one year was £40,000.

In both cases the destructive insect *Liparis monacha* was the cause of injury.

CHAPTER XXII

FUNGUS GROWTH ON TREES

SOME fungi attack living wood, some dead, and some both; and, while the majority are found growing on the stem, others attack the twigs, leaves or root. They are of all sizes, from the almost microscopical to the beefsteak fungus, which often exceeds a couple of feet in diameter. Some are jelly-like in consistency, others leathery or almost woody in texture; and while certain kinds spring up and disappear in a few days, others remain intact on the tree stem for several years. In shape, too, they vary considerably, from the curious Jew's ear fungus, which greatly resembles the human organ, to the giant puff-ball, that sometimes measures four feet in circumference. Then the cup-shaped and parasol fungi are appropriately named, while the flat, shelf-like arrangement of a species that frequents the oak and elm has interested many a casual observer. Excepting perhaps green, they are found of almost all colours from dull brown to the brightest crimson, while beautifully spotted and mottled kinds are not uncommon.

Healthy trees growing under normal conditions are little affected by fungi, whereas such as are predisposed to infection, by unsuitable soil, excessive drought or dampness, injury by accident or otherwise—in fact, those in a weakened condition however brought about—are, of course, far more liable to attacks. Though certain species of fungi only attack dead and decomposing wood, and are, therefore, the consequence and not the cause of disease, there are others, fortunately fewer in number, that attack healthy living trees and are the direct cause of disease and death. The prevention and extermination of fungus pests is by no means an easy matter, but by keeping the woods free from dead and dying trees and branches, and by giving immediate attention to outbreaks of the malady, much harm may be averted.

LARCH CANKER

In the following notes, only the most serious cases are dealt with, or, in other words, such fungi referred to as are commonly met with and cause most damage to our woodland trees.

Larch Disease.—In all probability the larch disease or

larch canker owes its origin to the minute spores formed in the fructification of the now well-known fungus—*Peziza Willkommii.* It has been pretty conclusively proved that these spores can only find a footing where the rind of the bark has become in some way injured, such as might be occasioned by the puncture of an insect, by wind, frost or from many other causes. The spores send down their germ tubes into the cambium, between the bark and the wood, where the moisture and nourishment afforded causes rapid development of the fungus. This soon spreads to the cells of the wood, and the annual layers either entirely cease to grow, or become disorganized and crippled in growth, causing a hollow appearance of the stem at the point of attack. The surrounding bark, by its attempts to heal over the wound, causes a thickened or burly appearance of the trunk, thus imparting to affected trees the cankered, swollen and distorted look that is so distinguished a characteristic.

The disease appears in this country on the larch, both common and Tyrolese, at all stages of growth up to thirty years, but rarely after that age. I have examined a plantation of only four years' growth sadly infested by the *Peziza* whereas, in other cases, the trees may be fully twenty years old before being attacked.

Cause.—Under what conditions of growth the larch is most susceptible to the *Peziza* is still a matter of vague uncertainty, but there can be little doubt that an enfeebled constitution, as fully explained in the article on the larch (" Trees for Economic Planting "), aided by our peculiarly erratic climate, has much to answer for. The variableness of our spring weather is, no doubt, one of the predisposing causes of disease, for, although no degree of frost experienced in this country can injure the tree when leafless, yet few are more sensitive when in young foliage.

Bearing on the subject of the larch disease, I have communications from almost every part of the country, and have personally visited and examined many of the worst infested plantations, particularly in England and Wales. Soil, if we exclude peaty, would seem to have little or nothing to do with encouraging the disease, as I have found it equally virulent on dry, sandy and heavy damp soils, but worse on chalk. Curiously enough, the disease is hardly known in the peat bogs of Ireland, and there the rainfall is excessive. It is equally strange that it first made its appearance in the Eastern and drier parts of England, and gradually proceeded to the more humid West. I have, however, noticed it in certain low-lying still, and damp portions of some woodlands, and where frosts would be most prevalent, but such cases are not general.

Remedy.—Several remedies have been more or less successfully tried with a view to getting rid of the disease on trees, such as by cutting and scraping out the injured portions, and applying a suitable dressing; but such remedies, although suited for single specimens, cannot be applied to a whole area of infested trees.

Under exceptionally favourable conditions, I have found the larch to outgrow the disease, though the cankered, swollen stems are never afterwards of great value for constructive purposes. Prevention in the present case is undoubtedly the best measure, for when once it has made headway, the larch disease is most difficult to cope with. By planting only sound, healthy trees, with uninjured roots in the most suitable soils and situations, and retaining these in as healthy a condition as possible, can we guard against attacks of the disease. Hardwoods mixed with the larch are to be recommended.

Polyporus squamosus attacks the elm and other hard-wooded trees and rapidly brings about decay. It is one of our commonest species on diseased trees.

The **Sycamore Fungus** (*Rhytisma acerinum*).—Every one interested in trees must be familiar with the conspicuous black, pitch-like spots which so mar the appearance of sycamore leaves. These are due to the above-named fungus which, appearing as small yellowish spots on the undersides of the leaves towards the end of June, gradually increase in size and intensity of colour until they attain to fully half an inch in diameter and are inky black, with a margin of dirty yellow. The attacked portion of the leaf becomes wrinkled and much thickened in texture, while all the affected foliage drops off prematurely. After lying on the ground during the winter, the thread-like spores are produced in large quantities at the time when the foliage of the sycamore is appearing in May and June. The sycamore is not the only tree affected by this fungus, for the Norway Maple and our native *Acer campestre* are equally

POPYPORUS SQUAMOSUS (ON ELM TREE)

liable to attack, and have in some instances suffered very severely from this cause. The fungus is becoming more plentiful, and it disfigures the maple and sycamore leaves to a wide extent.

Rhytisma punctata may at once be distinguished from the above species by the many small black spots studded thickly together, which combine to form the large conspicuous blotches for which affected trees are remarkable. Both species are sometimes found on the same leaf. As the attacks of this fungus continue from year to year, and cause almost every leaf to drop off prematurely, the health of the trees is greatly impaired, and they become an easy prey to the still more destructive coral spot fungus, *Nectria cinnabarina*. By burning the leaves affected with *Rhytisma* before the spores are liberated in spring, the spread of the fungus is prevented in a simple and effective manner. The sycamore fungus is very plentiful on trees around London.

The canker of hard-wooded trees is brought about by one of the Nectria (*N. ditissima*), and is very common in every part of the country, affecting the oak, beech, ash, sycamore and fruit trees generally. The ash perhaps suffers most, the timber turning black and being quite unfitted for structural purposes. It is strictly parasitical, growing on wounded portions of the stem and branch, and spreads with great rapidity, attacking trees of all ages. We have known the trees in a plantation of ash to be quite destroyed by this canker, which attacks most freely those growing on wet, sour land. This should be a warning to planters to avoid such soil.

The **Red-rot Fungus** (*Fomes annosus* or *Trametes radiciperda*) attacks the roots of several species of pine, particularly the Scotch, Corsican and Weymouth, as also the Spruce and Silver Fir. It is probably the most destructive of the family, attacking living roots and spreading rapidly from tree to tree. The trees, when affected, quickly turn sickly and die, the wood becoming spongy and of a brownish colour, with distinct black spots. The only remedy is to take out affected trees and burn them root and branch, replanting the ground with beech or elm. Though one of the most destructive fungi in coniferous woodlands, its attacks are by no means confined to these alone, since it is found on the roots of several hard-wooded species, such as the filbert, hazel, birch and beech. Quite recently a nut plantation in Kent suffered severely from the attacks of this fungus, whole lines of trees being killed outright before the cause was detected. It spreads quickly underground from tree to tree, and unless eradicated, as it may easily be by uprooting affected trees and carefully destroying the mycelium, much damage may result.

The **White-rot Fungus** (*Fomes igniarius*) is usually found on hard-wooded trees, and takes possession of wounds that may have been occasioned by the accidental breaking of a branch or stem injury. Oak trees suffer most from the attacks of this fungus, which is often as large as a foot across, and of a brownish colour. It is named igniarius on account of the inner surface being used as tinder, when prepared with saltpetre, especially in the old days of flint and steel. By removing the fungus and attending to the wound by cleaning away all dead and dying matter, and coating with tar, much benefit will ensue.

The **Leaf-shedding Fungus** or **Pine-leaf Scurf** (*Lophodermium pinastre*) is a well-known and widely distributed species, and is usually found on the Scotch and Austrian Pines. The leaves, when attacked, wither suddenly and fall off, the fungus being most prevalent after unusually dry weather, or in early spring succeeded by a frosty winter. It occurs both as a saprophyte on dead pine leaves and a parasite on the living foliage. We have been most successful in combating the attacks and preventing the spread of the pest by spraying the affected parts with "Bordeaux mixture." Young trees under ten years of age are most commonly attacked, and when this occurs in the nursery borders, the plants should be rooted out and burned. On several Scottish and English estates, thousands of Scotch Pines have been killed out by the attacks of the leaf-shedding fungus.

One of the commonest fungi on old stumps is **Agaricus melleus.** It has no particular host, being found alike on coniferous and hard-wooded trees, and on both root and stem. Known as the honey fungus, and edible, this toadstool is about three inches in diameter, and of a yellowish-brown or rusty colour. It spreads with great rapidity both in the soil and between the bark and wood of the affected tree. The only remedy is digging out and destroying the fungus, and, in the case of healthy young trees, collecting and burning the mycelium.

Next to the larch canker, one of the most destructive diseases of forest trees is caused by attacks of the **Bladder-rust** or **Cluster-cup** (*Peridermium piniacicola*). It is a wound fungus and attacks almost every species of pine, the

POLYPORUS SULPHUREUS (ON WHITE POPLAR)

Scotch in particular, especially when the trees are growing on light, poor soils. Young trees up to, say, twenty years old are most commonly attacked by this fungus, which appears like blisters, emitting bright reddish-coloured spores. Rooting up and burning all affected trees is the best remedy.

Polyporus sulphureus attacks many species of trees—yew, poplar, etc.

CHAPTER XXIII

BARKING OAK

Previous to the war the price of oak bark—about £2 10*s*. per ton—rendered it questionable whether from a purely financial point of view the operations of stripping and harvesting should be undertaken. Now, however, with bark at fully three times that figure, the operation is to be recommended.

The period of bark-stripping and harvesting is one of the most anxious seasons of the year with the forester, as the quality of the bark is so largely dependent upon the weather during the time that intervenes between the stripping and stacking, or delivery, and not less so upon the carrying out of the work at the proper time, to secure easy and expeditious peeling. In most cases the time when the bud is just expanding into leaf is that which gives the greatest weight of bark of the best quality, with the smallest amount of labour. By deferring the work, even for a few days, there is often a loss in weight amounting to as much as 10 per cent., and a great deterioration in quality.

Even in the most favoured situations it is seldom that the season for stripping extends beyond twenty-eight days. The advantages of early stripping are so well known that any comment on the subject is unnecessary; suffice it to say that, immediately the bark commences to "run" freely, no time should be lost in making a start, and the work should be prosecuted with vigour and dispatch until completed.

The proper time to commence barking cannot, however, be fixed with any amount of certainty, much depending on the season, whether early or late, as well as on the district of the country in which the operation is to be performed. During ordinary seasons, and in most parts of England, bark-stripping commences during the third week in April and continues for about a month, or until such time as the trees are in full leafage, whereas in some parts of Scotland, especially the north, the operation is frequently nearly a month later. No mistake can, however, arise as to the right time to start barking in any locality, as in all cases the period when the bud is first bursting into leaf will be found the proper time for felling to insure easy stripping and the best quality of bark. As the season of bark-stripping is, therefore, of short duration, every preparation should be made beforehand—trees marked and numbered, tools in readiness, and squads arranged—so that an early start may be made, as, by deferring the work beyond the time stated above, there is not only a perceptible loss in weight, but considerable deterioration in the quality of bark as well.

Elaborate directions regarding the arrangements of squads and tools to be used are unnecessary, as almost every district has its own peculiarities in this way. The tools generally in use are heavy axes and the cross-cut saw for felling, hand-bills and saws for pruning, peeling-irons or chisels for removing the bark, scrapers for removing moss, and light wooden mallets for beating refractory bark or such as cannot be removed by the peeling-irons alone.

Previously to felling the trees a man or stout lad is sent before, who removes the bark from the root upwards for a distance of 2 ft. or 3 ft.; this not only prevents its being injured when laying in and felling the tree, but is convenient for after-stripping as well. When the stools are intended for reproduction great care is necessary to avoid tearing or loosening the bark from the roots. After being thus prepared the trees are felled in the usual manner, those under 6 in. in diameter being cut with the axe; above that size it is found an economy of time and timber to fell with the cross-cut saw. Following in the rear of the cutters should be a squad of men, to clear the trunk and larger limbs of all branches down to 1 in. in diameter, leaving the limbs to be peeled as part of the tree.

Heavy timber and large branches are usually peeled where they fall, but it will be found convenient to have the smaller trees and branches carried out to some open space adjoining the stacking ground, and peeled while one end is supported by means of two forked sticks placed against each other. When the bark of small branches cannot readily be removed by the peeling-iron, a smooth and flat stone is brought into use, beside which the peeler sits, and with one hand holds the branch on the stone, moving it along from one end to the other, at the same time applying

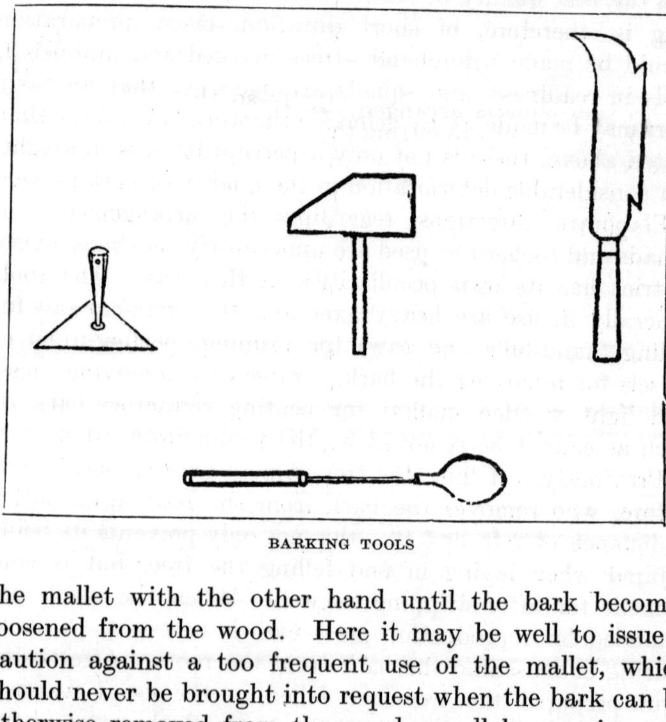

BARKING TOOLS

the mallet with the other hand until the bark becomes loosened from the wood. Here it may be well to issue a caution against a too frequent use of the mallet, which should never be brought into request when the bark can be otherwise removed from the wood, as all hammering and beating not only diminishes the quality of tannin, but has a tendency to blacken the fleshy part of the bark and cause rapid decay in a bad season. The body, or trunk bark, is removed in lengths of from 30 in. to 36 in., and in as large pieces as possible.

A dry, open and airy situation, convenient to the work, but without the wood, should be selected on which to harvest the bark, and rather than this should be done in a sheltered, humid spot, the bark should be carted to some distance off. The drying racks, or ranges, may be fully 2 ft. high, drooping somewhat to one side, and formed of forked sticks driven firmly in the ground, while stout rods are placed transversely upon these. It should also be so arranged as that not only may the rain be thrown off, but so that the ends of the bark may be facing the prevailing wind, thereby insuring a current of air through and beneath the mass. After being carted or carried to the drying-ground, the small bark is spread out loosely on the stage to a depth of about 6 in., and thatched or covered over with the larger pieces as a means of protection against rain. Each day's bark should be cleared up, and put on the range the same evening, and oftener, if found necessary, during damp showery weather, keeping the white or fleshy part downwards and using the larger pieces as covers to run off the rain.

During favourable weather the bark will be ready for stacking in about a fortnight from the time it was placed on the stage, but should close damp weather intervene, it may be found necessary to turn the bark occasionally, thus adding to the length of time required for harvesting. It should, however, be remembered that the less turning the bark receives after being placed on the stage the better will the quality be. Well-seasoned bark has the fleshy side of a creamy colour, whereas such as has been exposed to the sun or rain is of a dull brown, and is wanting in tannin matter, and consequently of inferior value.

As soon as the bark is thoroughly dry and ready for stacking, which may readily be ascertained by its breaking freely across rather than bending or yielding to pressure, it should be secured in a shed, ricked, or delivered to the tanner.

In stacking bark the rick should not be made too wide—say about 9 ft.—but well hearted, so that the side pieces may have a good fall or declivity outwards to throw off the rain. The rick may be of any length, according to the quantity of bark on hand, and of a height proportionate to the width. The largest pieces of bark should be reserved for thatching the rick, the whole being covered over by a tarpaulin or waterproof cloth of some kind. In most cases the bark is chipped previously to being sold, but as this necessitates having a large shed at command, the system is not generally adopted. There are, however, several advantages accruing from this method, not the least of which is that the bark may be chopped up as it is removed from the drying-stands, thus saving the expense of stacking. Chopping the bark can also be done by the workmen during wet weather, and when not otherwise engaged.

In computing the quantity of bark before stripping, we have found the following data fairly reliable :—

1. A well-balanced tree with a good head will yield about 6 cwt. of bark for every ton of measurable timber, if branches down to an inch in diameter are peeled.

2. Hedgerow trees usually yield about a ton of bark to every three tons of timber.

3. Trees growing in close woodland are usually thin barked, the yield being about a ton of bark to every 4½ tons of timber.

4. Oak poles will average five tons of timber to a ton of bark.

Tall, clean stems, as are produced when the poles are grown thickly together, with small heads, give the smallest yield in proportion to the quantity of timber, and short stems with spreading heads the largest.

The cost of production is as follows :—

	£	s.	d.
Labour, peeling and harvesting . . .	1	10	0
Cartage to railway station, including loading (this is the average from six districts) .	0	7	6
Loss on four months' delay in selling the timber, which, but for the bark, would have been felled in winter at 5 per cent.	0	5	0
Customary terms of payment, less 2½ per cent.	0	2	0
Superintendence, etc.	0	1	6
	£2	6	0

It has been carefully estimated that there is a loss of fully 12 per cent. of wood, caused by felling the oak during the barking season. In other words, the proportion of sapwood to the whole tree is about 14 per cent., which, for the majority of purposes to which oak timber is applied, is rendered by the peeling of little or no value.

We may say, however, that 2 per cent. can be profitably utilized, which still leaves us with a considerable loss on the whole tree.

The average price of the best oak is 2*s.* per ft., 12 per cent. of which is as nearly as possible 3*d.* per ft., or £1 17*s.* 6*d.* per ton of bark to 150 ft. of wood.

This, with the £2 6*s.* per ton cost of production, brings the total to £4 3*s.* 6*d.*, leaving a considerable balance on the wrong side at pre-war prices.

CHAPTER XXIV

THE MANUFACTURE OF CHARCOAL

AMONGST dead or dying industries of our woodlands that have been revived by the war, none is at present receiving a greater share of attention than the manufacture of charcoal.

There was a time, and not so long ago, when the merry voice and ring of the charcoal burner's axe were familiar sounds in some of the Kentish and other forests of Southern England; but keen foreign competition, aided by preferential carriage rates, have caused this time-honoured industry to slip from our hands; indeed, it was almost forgotten till again called into existence for the battlefields of France and Flanders. The trenches must be warmed without apprising the enemy of the existence of our men, and in order to do this and prevent soaring signals of smoke, the tent brazier is filled with glowing charcoal.

Except, perhaps, in Kent and Surrey and the English Lake district, where small quantities of charcoal are annually produced for the hop kilns and iron smelting, charcoal burning is a thing of the past. The expert charcoal burner is now a difficult man to find, and an independent, highly-paid workman when you have found him. Successive members of the same family in Kent have been known to follow the occupation of charcoal burning for fully a century and a half, and it is distinctly a skilled industry, and confined to few.

Usually the men work in threes, and, having selected a piece of ground sheltered from the prevailing winds and in a position to which easy access with wood can be obtained, a rough hut is erected for the accommodation of these nocturnal workmen. Water, sand or sawdust and turf are other requisites that must be provided as the work proceeds. A couple of large tarpaulins and half a dozen straw-covered hurdles are other necessities.

From the point of economy in carting the wood to the kilns it may seem that shifting the position of burning from one part of the woodland to another is to be recommended. Such is, however, not the case, as the hard, dry, ash-covered site, where charring has already been carried out, has its advantages, and the cost of transferring the workmen's hut and tools from one position to another must also be considered.

Several methods, largely dependent on the quantity and quality of charcoal to be obtained, are adopted, but in order to procure that of the best description the following system, which has been successfully carried out on a large estate for the past hundred years at least, is recommended. The timber carted to the charcoal yard consists of all kinds of hardwoods, preferably not under two inches in diameter.

Firewood and rough, unsaleable timber, as also inferior grades of heavy coppice wood, are mainly utilized for the production of charcoal. The wood is sawn into pieces about 2 ft. long, this being the most convenient size for building the kiln, and these again split if required to some 4 in. to the side, and when a sufficient quantity for two pits has been cut up, the building of these is proceeded with. It has been found economical to burn two pits at the same time, as both can be attended to as conveniently as one, and it is unnecessary for the men to sit up at night to watch each separately. The charcoal pits, one of which is shown in the accompanying sketch, are made of a broadly conical shape, 21 ft. in diameter and about 9 ft. high, and the mode of construction is as follows :—

A strong stake is driven firmly into the ground and left protruding about a foot. Around this are placed small pieces of dry ash of equal length, and standing as close to the upright stake as possible ; around this another layer is placed in the same manner, and this is continued until a circle 5 ft. in diameter is obtained. A circle 1 ft. in diameter, and having the top of the stake previously driven into the ground as centre, is next made by placing the wood horizontally on the upright pieces and side by side, the ends of each piece being placed at the circumference of the circle already made, and directed towards its centre. Layer upon layer is built in this manner until the pit is of the required height, the wood used here being dry pieces of ash 2 ft. in length, but split rather smaller than the ordinary pieces. A sort of chimney is thus formed, by means of which the pit is fired. Outside the core the wood is placed on end and reclining inwards, this being continued until the pits are of the required size. When the building is completed the pits are covered with newly cut turf, the grassy side placed innermost, beginning at the base and working towards the top, each line of turf overlapping the previous one by a few inches. The circular hole or chimney is left open for firing. Before turfing the top half of each pit it is carefully examined, and any crevices between the wood packed full of small pieces of turf and sawdust to exclude the air. The turfs are cut about 1 ft. in width, and of any convenient length. The quantity required for two pits of the dimensions stated is seven loads.

When the pit is satisfactorily covered it is fired by dropping a couple of shovelfuls of burning wood and some dry pieces of pine or ash into the opening left at the top ; the top turf is then put on, which effectually shuts up the chimney, and the process of charring commences. The smoke is first seen issuing from the lower half of each pit, where the chinks were not packed with sawdust, and ultimately it escapes from the whole surface.

Constant attention is required day and night during the period of burning, especially should the weather be stormy, as the wind, by striking on a particular part of the pit, causes that side to burn more rapidly, and fall in. When this occurs the hole must at once be filled in with rough logs, which had been set aside for the purpose when splitting the wood, and re-covered with turf.

When the weather is mild the pits burn uniformly, require but little attention, and produce the finest charcoal. The time required for burning will vary with the size of the pit, quality of wood, method of covering, and meteorological conditions. From six to seven days are usually required for pits of the above dimensions, but smaller kilns only covered with grass, fern and a little soil may be ready for uncovering in from two to four days. Long experience has, however, proved that by the slower process of charring the best charcoal is produced, but the cost is higher. By covering the pits with grass and fern, as is often done, a considerable saving is no doubt effected, but where turf is available there can be no question as to its value over the former, and on the boundaries of most woodlands it is readily procurable at the cost of cutting. As the charring proceeds the turf gradually disappears until only a slight covering of burnt earth remains. When the pits have burned out and become cool, it is found that they are reduced to rather less than half their original size.

SECTION OF CHARCOAL PIT

The charcoal is extracted by means of a specially constructed rake resembling a light drag, but having much finer teeth, which, after it has become quite cold, is stored in a shed until required for use.

The very finest charcoal, superior to what is generally sold, is produced by this method. The expenses connected with making it are, however, a little heavier than usual, owing to the slower system of charring, the use of larger wood, and the extra cost of covering with turf. As to the cost of producing charcoal by the above method, this will vary greatly, much depending on the distance the wood has to be carted and on the cost of labour in the particular district.

The price paid to the charcoal burners is 7*d.* per bushel, or about four guineas per ton, which may seem high, but when we consider that it is specialized work that is confined to few and attended with grave risks and discomfort, the amount earned is not excessive. It should also be remembered that, previous to lighting the kilns, sufficient rough, not corded, wood has to be sawn and split and the pits carefully built and covered, not to speak of the constant attention required, both day and night, wet or dry, for from three to seven days, during charring process. The usual price for burning charcoal when the wood is corded is 35*s.* per ton.

Fresh-felled wood is rarely converted into charcoal, the greater portion of that used being thinnings of the previous season. The proportion of wood to charcoal varies greatly, much depending on the size, quality, and maturity of timber. Having had occasion to purchase charcoal lately, I found the price, retail, to be 2*s.* 6*d.* per bushel, or in quantities of not less than a ton, £14, for that of fair quality.

From about the twelfth century onwards Scotland, where wood was abundant, produced annually a large quantity of charcoal iron; and in 1660 the Navy Commissioners nominated John Evelyn to investigate the then denudation of forests owing to the manufacture of charcoal for iron smelting, and the following quaint extract from his report will be interesting :—" Nature has thought fit to produce this wasting ore more plentifully in woodlands than any other point, and to enrich our forests to their own destruction —a deep execration of iron mills and ironmasters also." The Lorn Works, in Argyllshire, were started in 1753, and annually consumed upwards of 3,000 tons of lump charcoal.

The Sussex and Kentish forests at one time supported many of the familiar charcoal burners, and right brawny and thrifty were these denizens of the woodland with their rustic wooden huts and piles of rifted firewood, but the industry was almost a thing of the past till again called into active existence by the exigencies of the war.

Kiln-burning.—The kiln is made of brick, one course being sufficient, if bands of iron be added to strengthen the brickwork. It is usually conical in shape, 24 ft. in diameter, with an equal height, and holds about forty cords of wood. The wall of the kiln is carried up nearly straight for about 6 ft., when it is gradually drawn in and made a blunt cone shape. A plate of iron is fastened on the top in the manner of a stone to an arch. Three-inch hoop-iron bands, about an eighth of an inch thick, are placed around the kiln and drawn together by means of screw-bolts and nuts. At the base, and near the top, are double sheet-iron doors, by which it is filled with wood or emptied of charcoal. The time required to fill, burn and empty is about three weeks. Pit-burning, for estate purposes, is, however, most commonly pursued, and has this advantage—that the charcoal can be made at any place where timber is being felled, without extra expense, save that of the cartage of the charcoal, whereas in using the kiln or retort the wood must, in most cases, be conveyed to the place where it is erected.

Comparative Value of Woods for Charcoal-making.—Amongst home-grown timber, oak, ash, and beech are generally preferred for charcoal making, but the following table will show pretty correctly the proportionate relative values of the various descriptions of wood for gunpowder charcoal :—

	Per Cent.
Rhamnus frangula contains	27
Laburnum	25
Boxwood	24
Sweet Chestnut	23
Oak	22
Holly	20
Walnut	20
Beech	19
Sycamore	19
Elm	19
Willow	18

	Per Cent.
Poplar	18
Birch	17
Alder	17
Ash	17
Hazel	17
Mountain Ash	17
Scotch fir	16
Larch	16

Uses of Charcoal.—The uses of charcoal for estate purposes are very numerous, for horticultural, agricultural and other departments. From remote antiquity charcoal has been used as a fuel, and for many purposes it is still unsurpassed. It is by far the cleanest solid fuel known; it burns steadily, gives out a great amount of heat, and lasts well. On account of its smokelessness it is invaluable for cookery, and it is also admirably suited for use in greenhouse and other stoves. It is not adopted for heating apartments on account of the poisonous gas (carbonic oxide) produced in its combustion, and the danger, most apparent when the charcoal is burnt in an open chauffer, is not obviated by using it in a stove, as carbonic oxide has the power of diffusing through red-hot iron.

In gardening, charcoal is largely used for potting purposes, for vine borders, and for flower beds; and in the form of dust it is the best material for packing bulbs for transmission to a distance.

Perhaps the most important of the uses to which charcoal can be put about a house or estate is that depending on its extraordinary power of absorbing gases. It is a perfect deodorant, a preservative of food and all animal substances and a valuable disinfectant. The gases most readily absorbed by charcoal are those which are most prejudicial to health and most frequently produced by putrefactive changes.

In the pores of the charcoal they are destroyed by union with the oxygen condensed from the air. The fact of its being absolutely non-poisonous and perfectly odourless puts it before all other disinfectants.

"Dogwood" for Gunpowder Charcoal.—The alder

buckthorn, berry-bearing or black alder (*Rhamnus frangula*) is a native shrub that is fairly plentiful in Southern England, though rare in Scotland and Ireland. Confusion sometimes arises from the same popular name being applied to widely different species of plants, and this, unfortunately, is the case with the shrub in question. What is known among gunpowder manufacturers as dogwood is in reality the present shrub (*Rhamnus*), which, however, is quite distinct from the true dogwood (*Cornus*) and belongs to an entirely different family. To those who contemplate growing charcoal wood for the making of explosives, this distinction is of the utmost importance, as I have seen *Cornus sanguinea* cultivated for the making of gunpowder.

The alder buckthorn is perfectly hardy, growing freely even in the North of Scotland, where it ripens its seeds. It is usually found as an erect-growing bush from 8 ft. to 10 ft. in height, though in suitable situations in Southern England specimens fully 20 ft. high, with stems 6 in. in diameter, are to be met with. The bright green leaves are oval in shape and vary, according to conditions of growth, from 2 in. to 3 in. in length, while the flowers are of a dull yellowish green and are succeeded by dark purple berries each about the size of a pea. From a very early date the alder buckthorn has been cultivated, though not extensively, in this country for charcoal making, and the price, upwards of £15 per ton, that is paid for the wood, shows that the growing of this shrub is a profitable undertaking. At one time large quantities of the wood were produced in Sussex and other counties, the selling price being £14 per ton when peeled and tied in bundles.

The cultivation of the alder buckthorn is nearly similar to that of the osier for basket making, and the produce is dealt with and disposed of in like manner. For soil any good loam inclined to be dampish will suit it well, and an open, but not wind-swept, situation should be chosen for its cultivation. The land intended for growing the alder buckthorn should be trenched the winter before planting, and a top-dressing of leaf-soil or thoroughly decomposed manure, the former preferably, will greatly assist the growth of the young plants and prevent too speedy evaporation of moisture from the soil. Young plants are not offered in quantity in our nursery catalogues, and in order to obtain a stock sufficient to form a plantation, seed-sowing or layering old plants must be resorted to. Fortunately, by either method the plant is readily obtained in quantity, and as the seeds are produced in fair abundance and ripen freely this method of getting up a stock is to be recommended.

The berries, after being collected in the early winter, are treated much as we treat those of the yew and holly. They are mixed with sand in order to separate the seed and fleshy covering, and the whole is sown during early spring in previously prepared beds. The seed beds may be prepared in any shady situation out of doors, the soil being largely composed of light sandy loam mixed with finely riddled leaf-mould. Sometimes the seeds are sown in boxes and placed in a cool frame, but we have found cultivation out of doors more satisfactory. When two years old, the seedlings should be transplanted into lines 18 in. apart and 9 in. from plant to plant. Here they may remain for another two years, after which they should be planted out permanently and headed back the following season. Rather thick final planting is to be recommended, as the shrub being of upright growth, requires comparatively small room for development, and the best wands are produced by a close order of growth, say 5 ft. from plant to plant.

Layering does not produce such upright-habited shrubs as those grown from seed, and the yield of wood per acre under exactly similar conditions of growth is greatly in favour of seedlings.

After planting, the ground should be kept free from rough-growing seeds for the first two years, the crop being cut at from six to seven years' growth, when the wands are from 1½ in. to 2 in. diameter at butt end. Cutting and bundling is usually done by contract, but, as with the osier, it is imperative that the crop be cut over near ground level and short " stumps " without " spurs " encouraged.

As in the case of ordinary coppice wood, the buckthorn for charcoal making may be cut every sixth or seventh year, the straightest shoots when sorted in about 5 ft. lengths being tied in bundles which are about a yard in girth. The buckthorn being a gross feeder, manuring the land after the removal of a crop has been found advantageous.

Although largely imported from Holland, and other parts of the Continent, home-grown wood is preferred, as it produces a much superior charcoal for the manufacture of explosives. Unfortunately, however, home supplies are so limited that foreign wood is imported in considerable quantity, and as there was a scarcity before the war, the probabilities are that, with our greatly increased consumption of charcoal explosives, a dearth of suitable wood is now being felt.

With the present small remuneration attaching to the cultivation of coppice or underwood, which under ordinary circumstances does not exceed £4 per acre, the cultivation of the alder buckthorn is to be recommended, particularly as the price is at least quadrupled. The cultivation of this is quite simple, while the quality of soil required need not be better than that which produces a crop of chestnut or hazel. In cultivating the alder buckthorn for charcoal purposes the following rules should be observed :—

1. It will not succeed satisfactorily in sandy, poor or water-logged soils ; rich, well-manured loam being preferred.

2. The ground should either be trenched or ploughed and cleared of all rough-growing weeds the winter before planting.

3. Plant seedlings or layers in the spring in lines about 5 ft. apart and the same distance from plant to plant.

4. An annual clearance of weeds and loosening of the soil between the rows of plants is recommended where a heavy crop is expected.

5. Induce the growth of stout clean shoots by liberal feeding and clean cultivation.

6. Cut the shoots close to the ground so as to prevent the formation of long spurs and minimize the number of offshoots. Clean cutting with a sharp tool is imperative.

7. After the removal of a crop, stirring and enriching the soil is to be recommended.

8. Though found mixed with undershrubs and in the shade of trees when in a wild state, yet the greatest quantity of the most valuable wood for charcoal making is produced in open situations.

CHAPTER XXV

PRICES OF HOME-GROWN TIMBER

These vary so much in different parts of the country that it is almost impossible to give a list that would apply generally to the British Isles. Local demand and the situation of the plantations where the timber is to be sold have to be considered, while the quantity available and quality of the particular timber are also important factors in determining the price.

In many instances woods and plantations are far removed from road and rail, consequently the cost of delivering to the consuming centre is proportionately high, while in the vicinity of coal mines or manufacturing towns almost every class of timber will find a ready and profitable market. Then the quality of the timber (its reputation, we might almost say) is a powerful factor in assessing its value, as will be seen in the case of oak produced in Surrey and Sussex, or beech from the chalky districts of Kent and Hertfordshire. No timber merchant thinks twice as to whether he should, on account of quality, purchase oaks at Rockingham, Ashridge, Welbeck or other well-known centres, and so it is with the beech timber on the Chiltern Hills, at the Chenies, in Hertfordshire, or with Scotch pine from the famous Aberdeenshire plantations.

But there is another factor that frequently affects the price of good timber, and that is, that in inaccessible positions the quantity offered is too small. This has been the case with excellent larch timber in Ireland, where the quantity offered was not sufficient to induce the timber merchant to lay down plant and arrange for delivery to the nearest railway or port. There are other causes of a minor nature that greatly affect the price of home-grown timber, and it is a curious fact that in adjoining counties, sometimes even in the same county, the prices of various timbers vary several pence per cubic foot. With all woodland products the same is the case, and locality and local demand determine the price of faggots, firewood and other commodities.

Previous to the war, the following were the average prices of home-grown timber, felled and lying in the wood; also of other woodland produce. The prices at present are generally much higher.

		s. d.		*s. d.*	
Ash	from	1 6	to	2 6	per cubic foot.
Alder	,,	0 7	,,	0 10	,, ,,
Beech	,,	0 10	,,	1 4	,, ,,
Birch	,,	0 7	,,	0 10	,, ,,
Chestnut (Spanish)	,,	1 3	,,	1 6	,, ,,
,, (Horse)	,,	0 8		—	,, ,,
Cherry	,,	0 9	,,	1 2	,, ,,
Elm	,,	0 7	,,	1 6	,, ,,
Larch	,,	0 10	,,	1 4	,, ,,
Lime	,,	0 10	,,	1 6	,, ,,
Oak	,,	1 3	,,	2 6	,, ,,
Oak, brown	,,	3 6	,,	10 6	,, ,,

(but these trees are usually sold at so much for each)

Poplar	from	0 9	to	1 6	per cubic foot.
Scotch Pine	,,	0 5	,,	0 10	,, ,,
Spruce Fir	,,	0 5	,,	0 9	,, ,,
Willow	,,	1 6	,,	10 6	,, ,,
Walnut	,,	1 0	,,	2 6	,, ,,
Sycamore	,,	1 3	,,	2 6	,, ,,
Firewood	,,	5 0	per cart load.		
,,	,,	12 0	to	16 0	per cord.
Faggots (large)	,,	14 0	,,	21 0	,, 100.
,, (small)	,,	2 6	per 100.		
Oak bark	,,	52 0	,, ton.		
Charcoal	,,	1 0	,, bushel.		

Since the war commenced several kinds of timber have increased greatly in value, notably ash, poplar, Scotch pine and spruce fir. For the latter as much as 1*s.* 4*d.* per cubic foot was obtained in Bucks, while in several cases first-class ash timber realized upwards of 4*s.* per cubic foot.

Pitwood, also, has increased in price.

Prices of Timber per Ton Weight

In several parts of the country, particularly Ireland, timber is regularly sold by weight and the following prices were realized previous to the war:—

	s. *d.*		*s.* *d.*	
Ash and sycamore	—		18 0	per ton.
Beech	5 0	to	10 0	,,
Larch	12 0	,,	20 0	,,
Scotch pine, spruce and silver fir	—		8 0	,,
Spruce	8 0	,,	10 0	,,
Oak	—		15 0	,,
Elm and beech	—		10 0	,,
Hardwoods of pitwood size .	—		6 0	,,
Larch poles and pitwood (Wales)	9 0	,,	10 0	,,

That the heavy importation of foreign woods has had a prejudicial influence on the value of home-grown timbers is realized by all those who are engaged in the trade. The depreciation in price has, however, been so gradual that it is difficult to realize this unless by comparing the prices of to-day with those of, say, a hundred years ago. Several of these comparisons clearly indicate that oak, at least, has become much reduced in value, less so ash and elm, and the following list of prices obtained in 1807 for timber on two estates in Hants will serve to show that for trees of equal size the price to-day is far behind that of a century ago.

				s. *d.*	
Oak	averaging	9	cubic feet . .	4 0	per foot.
Ash	,,	7	,, . .	2 3	,,
Elm	,,	22	,, . .	1 9	,,
Beech	,,	14	,, . .	1 6	,,
Sycamore	,,	19	,, . .	1 3	,,
Fir	,,	11	,, . .	1 3	,,

On another estate trees of about the same size brought: oak, 4*s.*; ash, 1*s.* 6*d.*; elm, etc., 1*s.*; beech, 1*s.*; and firs, 1*s.* per cubic foot. When the small size of the individual trees is taken into account, and even admitting that only the best portion of each trunk was measured, the prices, as compared with those of to-day, are exceedingly high, particularly for oak, ash and fir.

CHAPTER XXVI

BRITISH TIMBER AND SOME OF ITS USES

THE following are a few of the many uses to which home-grown timber is applied :—

Alder is used extensively for clog soles, barrel staves, mill bobbins, and occasionally in furniture making. It makes excellent charcoal for cooking and heating, as well as that used in the manufacture of gunpowder.

The wood of the **Apple, Cherry** and **Pear tree,** when of large size, is used for cabinet purposes, and stained in imitation of other woods. For veneers, golf clubs, bowls, etc., these woods are of value, as also for weaving shuttles.

Ash timber is largely used by agricultural implement makers on account of its possessing great elasticity and bearing considerable cross-strain. It is the best wood for shafts of all kinds, for tool handles and wooden rakes, and is largely used by furniture makers.

Beech wood is the chief constituent of cane-bottomed chairs, and is largely employed for the handles of joiners', carpenters' and other wood-workers' tools. For gun-stocks, saddle-trees for heavy harness, wheel-felloes and bobbins it is also largely employed. When of large size and clean growth, it is used for calendar machines, and for engineering purposes in spinning and bleaching districts. It makes excellent charcoal.

Birch wood is largely used for turnery work, thread bobbins, clog soles, shoe pegs, furniture, hatters' blocks; it is also used in the manufacture of brushes and in toy making.

Chestnut (Spanish) timber more nearly approaches that of oak than any other species, and when stained is not only substituted for it, but for the walnut as well. For piano sides it is largely used, as also for rafters in open-roof churches, for furniture and cabinet work, ship fittings, sign-boards, and post and rail fencing.

Elm wood is extensively used for the boarding and flooring of carts and wagons, in coffin making, for the framework and foundations of bridges, for naves for wheels, and for the keels of boats and ships. It makes strong furniture, and is often substituted for ash in making agricultural implements.

Holly is used by mathematical instrument makers, for fancy turnery and inlaid work. It is often sold as ebony when "ebonized."

Hornbeam timber for cogs in mill gearing is well known, also in "bushing" for sawmill rollers, and for skittle pins.

Horse Chestnut.—The timber is largely used for packing boxes, moulding patterns for castings, cutting boards, manufacture of brushes, and occasionally for covering temporary buildings.

Larch.—The wood of this tree is largely used for fencing, boat building, permanent staging, and pit wood.

Lime.—The wood is white and very fine of grain, and used for carved work, sounding boards for musical instruments, wagon brakes, packing boxes, toys, domestic utensils, and for shoemakers' and saddlers' cutting boards. Charcoal for gunpowder is made from this wood.

Maple is employed in the turning of bowls, for toys, and "bird's-eye" maple for furniture.

Oak has long been associated with our national defence as the chief element in shipbuilding, but although iron and steel have to a great extent taken its place, yet for barges and small boats the timber is still largely used. Wagons for railway mineral traffic are largely made of oak, while the builder finds in it his best material for the strong frames of domes, spires and roofs of public buildings. It is also used for the bottoms of carts and wagons, cartwheel spokes, fencing, furniture making, railway "spraggs," charcoal, etc.

Poplar wood is woolly and tenacious, and for this reason is used for the bottoms of stone carts and barrows. It is well adapted for making packing cases, railway brakes, weather boarding, and for purposes where lightness is of greater importance than durability. The Abele, or white Poplar, produces perhaps the most valuable timber of any of the numerous species.

Scotch Spruce and **Silver Fir** may all be classed under the same heading, being of about equal value and applicable to similar purposes, viz., for sleepers and pit wood, boarding under slates, headings for barrels, soap boxes, temporary fencing, also for conversion into planking for lead works, and for all erections of a temporary kind.

Sycamore timber is peculiarly white and smooth and free from grain, which makes it very valuable. It is used for curtain rings, churns, butter prints, for the backs of violins, for founders' patterns and cutting boards, and in the making of wooden vessels and furniture. For calendar machines and in cotton and jute factories it is much employed.

Walnut timber is much in demand for gun and rifle stocks, for the best class of furniture, and for veneering purposes.

Willow is famous for the production of the best class of cricket bats and for artificial limbs and crutches. It also makes good charcoal.

Yew wood is valuable when employed for veneering.

CHAPTER XXVII

WILLOWS FOR BASKET-MAKING

Previous to the war, preferential railway and boat rates, aided by keen foreign competition, wellnigh rendered the time-honoured industry of basket-making a thing of the past in this country. The best classes of osiers, cleaned and ready for manipulation, were delivered to our principal markets from Continental sources at so low a price that competition on our part was almost out of the question. There are still, however, a few stations, such as those in Bedfordshire, the fen districts of Lincoln and Cambridge, and along certain reaches of the Thames, where willow culture is engaged in, though not in the same energetic way as was the case some half a century ago.

About 7,000 acres, producing roughly 20,000 tons of osiers, are cultivated in this country at the present time, many small plantations having been grubbed out and the land laid down in other crops during the past five and twenty years.

This falling off is much to be regretted, as the sorting and harvesting of osiers and basket-making gave light and remunerative employment to a large number of residents, both young and old, of the districts in which the willow-beds were situated. Land that was damp and could not well be brought under other cultivation without the expense of drainage, gave a good return under a crop of osiers. Taking everything into consideration, and judging from talks that I have had recently with those who are interested in the osier industry, the formation of willow-beds, under the plea that considerable profits attend the undertaking, is not to be recommended. No doubt in some favoured districts where carriage is reduced to a minimum and local demand is considerable, osier cultivation gives a fair return for capital invested; but until we can get back the once lucrative trade in baskets for fruit and other similar commodities from the hands of our Continental rivals, profits of any magnitude are quite precluded.

There is little doubt that on soil which is unsuited for farming purposes willow culture can be made fairly profitable, but it is a mistake to suppose that any marshy piece of stiff ground will grow osiers, or that the planting and tending are matters of small import. Quite the reverse is the case. The willow will not thrive for long in water-logged soils, though periodical inundations, particularly during winter and early spring, are highly beneficial. In addition, the soil must be well worked, the cutting of rods carefully and systematically carried out, and the clearing of the ground must receive strict attention, else deterioration of the crop will quickly ensue. Previously to planting the cuttings or sets, the ground should be ploughed, harrowed and consolidated, all objectionable weeds being destroyed.

For some considerable time to come there is bound to be a dearth of willows for basket-making, as our main supplies have come from Germany and the Netherlands. Here, then, is a chance for the owners of suitable land in this country to set to work at once and revive a time-honoured industry by planting up suitable grounds with the most approved kinds of willow for basket-making. That the undertaking, if wisely carried out, would be a remunerative one is beyond question, and the excellent results attained at such places as Leicester and Bedford clearly prove that willow culture is a most profitable way of utilizing naturally dampish land in any but the most exposed situations.

For the past few years the demand in this country for high-grade willows has been greatly in excess of the supply, in fact, hardly one-fifth of our requirements are produced at home. Germany, previous to the war, exported willows and rods to the value of about £42,000, this being an increase in five years of fully one-half; while of the manufactured articles in the way of baskets and basket ware her total value exceeded £42,000. But as giving some idea of our wants in this direction it may be stated that the total value of willow rods annually sent to this country from the Continent is in round figures about £100,000, and of baskets and basket ware fully £170,000. What a contrast with the period in our history when an important export trade in willows was done by this country!

The willow working industry is a rapidly expanding one. Owing to the increasing demand, the value of peeled willows is gradually on the increase, and present prices range from £24 to £38 per ton for those of best size and quality. These are in the main exported and used for high-class work in the basket trade, rougher unpeeled willows that are largely in use for cheap packing hampers and farm purposes bringing in a much lower price. Fruit baskets in immense numbers are annually imported from the Continent, one firm alone having sent over £150,000; while at Leith basket works, which mainly caters for the agricultural and fishing industries, thousands of herring baskets alone are sent out every month, while the packing hamper department is of great interest and a special feature of this enterprising firm.

Previous to the war willow or osier culture was mainly in the hands of the French for rods of good quality, the Belgians and Germans supplying a cheaper kind probably owing to the quality of soil and inferior varieties that are cultivated.

There are not a few persons who consider that in order to cultivate willows successfully, any neglected, damp piece of ground, which is unsuited for other crops, may be utilized, and the cuttings simply stuck in without ground preparation of any kind. This is, however, a great mistake, as experience has long ago demonstrated that in order to make osier cultivation at all profitable, a low level, and a naturally rather moist situation must be chosen, and further, that the soil should be deep, well drained and thoroughly prepared.

Thoroughly drain the ground first, then steam-plough or trench the soil to a depth of about 18 in., removing carefully all weeds, particularly such troublesome kinds as the bindweed, couch grass and dock. It is always preferable to take a crop of potatoes first from the land intended to be laid down for osier culture, as it not only sweetens and enriches the soil, but allows of the eradication of all obnoxious weeds. Where, however, it is not practicable to crop the land first with potatoes, the soil should be well and roughly broken up and left so for a year, or for a winter, at least, before being planted with the osiers. The best time to plant is from October to the middle of March. The sets, or cuttings, should be about 15 in. long, and formed of well-ripened rods, of one year's growth, and the straightest and cleanest portion of the rod only used. Three or four buds should, if possible, be on the top end of each set. In planting, insert the cuttings from 9 to 12 in. into the ground, leaving 3 in. above soil, which forms the stool that bears the future crops.

It is well to exercise great caution while inserting the cuttings, as, if the work has been delayed till the sap is rising, the bark readily strips away from the wood, and this is very objectionable, as the plants in such a state usually die. The sets may be placed about 15 in. apart, and the rows, which, for convenience, should be lined off straight, about 30 in. from each other. Of course, as regards distances, these will depend to a great extent on the quality of the soil and the particular kind of willow being planted, but the above are good average distances.

For basket-making, etc., the best kinds of osiers to use are *Salix triandra* and *S. viminalis*, but there are others. A good basket-willow, be it of whatever kind, should, when green, twist from end to end without breaking. It is well to bear in mind that, in order to obtain the greatest profits from willow culture, only the very best kinds should be planted—indeed, next to preparing the ground, a judicious selection should be one of the main considerations. For the first year, at least, after being formed, osier-beds must be carefully attended to in the way of cleaning and weeding. Hoeing will be found the most convenient method of getting rid of weeds, but, in the case of bindweed, handpicking around and amongst the sets will be found necessary.

Cutting the osiers must be done while the crop is dormant, or not later than the middle of February, but not during frost, which will injure and kill off parts of the stool. The cutting is done by means of a rod-hook, which resembles a miniature sickle; this should always be kept sharp, so that the cuts may be made clean. Tie the rods together when dry, in bundles of three or four sizes, and either house or stack them. It should be borne in mind that rods are easily spoiled by being tied up or stacked whilst in a wet state, as they very soon become heated, which makes them brittle and utterly valueless for the purpose intended. What is known in England as "bolting" is simply taking a number of osiers, as nearly of a size as possible, and laying them on a twisted wand, at the same time keeping the butts all one way, and level, then drawing them tightly together—not, however, to such an extent as to injure the bark—with a rope and two levers, and finishing off by tying the wand. The wand should be at 14 in. from the butts. A bolt of rods should measure 40 in. round the band.

In forming a willow bed, the following short rules should be observed:—

1. Willows will not succeed well in peaty, sandy, or water-logged soils; rich, well-drained loam, that can be flooded at will is the most suitable.
2. Trench or plough, and thoroughly clean or pulverize the ground before planting.
3. Plant only the best kinds, studying soil and market, and avoid a mixed crop.
4. From November to March insert the cuttings about nine inches deep, avoiding such as are bark-chafed, and tramp firmly.
5. Keep the beds clean and free from weeds.
6. Cut the crop close to the ground; pollard willows soon decay and in that state harbour injurious insects.

The following estimate of the approximate cost per acre of osier culture, and the returns therefrom for the first three years will be of interest:—

FIRST YEAR.

	£	s.	d.
Ploughing the ground and planting . .	4	10	0
Hoeing and other attention	1	5	0
15,700 willow cuttings (*Salix viminalis*) . .	10	0	0
Rent, rates, and 5 per cent. interest on capital	2	12	0
Harvesting	0	18	0
	£19	5	0
Yield first year 3 tons, value.	9	0	0
Loss	£10	5	0

SECOND YEAR.

	£	s.	d.
Rent, rates and incidentals	2	10	0
Hoeing and cleaning	1	8	0
Harvesting	1	10	0
	£5	8	0
Yield about 5 tons	20	0	0
Profit	£15	8	0

THIRD AND SUBSEQUENT YEARS.

Expenditure £6. Yield £24. Profit £18.

If the ground is properly cultivated and losses made good, the plantation should give a yield similar to the third year for fully a quarter of a century.

In the low-lying district between Taunton, Bridgwater and Langport, in Somerset, willow or osier culture is largely engaged in.

The system generally adopted is that the owner or tenant of the land planted to willows keeps the land free from weeds to prevent the withies being choked. This entails an outlay of about 25*s*. per acre per annum if properly done. The crops are measured and marked out in half-acre lots and sold in October or November. The purchasers cut and remove them, and in some cases convert them into baskets, basket-chairs, and such like. In other cases the purchasers select and bundle the crop in the regulation sizes and sell them to dealers.

The results of some recent sales are as follows: One field of $11\frac{1}{2}$ acres realized £132 and the first two half-acres, being remarkably good withies, made £19 10*s.* Three other fields, containing 16 acres in all, made £161 10*s.*; and three others, containing $22\frac{1}{2}$ acres, £222 2*s.* 6*d.* The agricultural annual rental value of this land when pasture, before it was planted to willows, was under £2 per acre.

From the above it will be seen that if prudently entered upon and economically carried out the cultivation of willows for basket-making is a paying industry, and as for some years to come foreign supplies will be barred to our markets, the enterprise should prove highly remunerative.

CHAPTER XXVIII

UTILIZING WASTE FOREST-PRODUCE

That the production of timber, in common with other trades, has of late years been rendered far less remunerative than formerly, owing principally to keen foreign competition, is a fact that is now well known, even to the most casual observer. In face of this it behoves us to ask ourselves the question: Do we utilize to the fullest extent the by-products of the forest and woodland, and so diminish waste, and, at the same time, add to the general revenue of the forest department? By waste produce, or by-products, is meant anything other than wood in the condition in which it is generally used, and includes bark, charcoal, firewood, house and kiln faggots, tar, wood-spirit, turpentine, sawdust, wood-ashes, leaf-soil, etc. That much may be done, both in economy of production and in utilization of waste produce, is well known to those in charge of woods and forests in every part of the country. Where the by-products cannot well be utilized in any of the above-named ways, it would be better, perhaps, to reduce them to ashes; for, by so doing, insect and fungus life are lessened, and a valuable manure is obtained, particularly rich in potash, whether for grassland or certain farm or garden crops.

The forest by-products of Great Britain and Ireland are, in the main, applied in the four following ways:—

(1) Firewood.

(2) Charcoal, for heating purposes, etc.

(3) Bark for tanning.

(4) Faggots, for house and kiln purposes.

Although the British forester has usually little or nothing to do with what we might term the volatile products of the forest—tar, pitch, turpentine, rosin, wood-spirit, acetic acid, etc.—nor, indeed, with paper-pulp, it may not be out of place to point out briefly the minor uses to which waste timber and by-products generally may be applied.

Tar, which at present is largely imported from the Baltic ports and Southern United States, is obtainable principally from three species of Pinus: *P. palustris*, *P. Pinaster*, and *P. sylvestris*.

Pitch is simply tar deprived of the volatile oils, which is brought about by boiling.

Turpentine comes from incisions made in the stems of some of the pines, principally *Pinus palustris*, *P. sylvestris*, and *P. tæda*.

The common silver fir (*Abies pectinata*) produces the famous Strasburg turpentine, while the larch is the source of the Venice turpentine of commerce.

In New England the whole of the younger sapling pines —stem, branches, bark and leaves—are made into pasteboard, while in other countries the lime and poplar are converted into paper-pulp of great value. That the great and ever-increasing demand for paper of all qualities will cause a corresponding demand for the material used in its production cannot be doubted, and attention has already been directed to this matter in some parts of this country. From the sap of the larch and Scotch firs "coniferin" is obtained, while "rubber," a valuable product for mixing with gutta-percha, which is very durable, is got from the bark of the common birch by distillation. The value of gorse as a food for horses and sheep is well known even in this country, while in Italy poplar leaves have long been used as cattle-food, and ground fir-needles in Styria for the same purpose.

Dried leaves make excellent litter, and they are valuable as manure. Sawdust, though without manurial value, absorbs liquid manure, and is thus used as an excellent top-dressing. Leaf-mould is well known for its many uses in the garden, as for top-dressing and mixing with other poorer soils in the making of composts for planting. These are some of the many uses to which the minor by-products of the forest can be applied, but, as these hardly come within the scope of the forester, special attention will be devoted to the major by-products—firewood, charcoal, bark and faggots—with which the British forester is most intimately associated.

(1) **Firewood.**—Never, perhaps, was the subject of English firewood more worthy of consideration than at the present time, when the price of other fuels is excessively high. Many persons will maintain that in districts where coal is abundant, it is very questionable whether there is any advantage to be obtained from burning wood. We have satisfied ourselves that even if wood could be procured at less than its present price—firewood price—it is nearly as expensive as coal, as sold previous to the war, in most of our large towns. No doubt, on many large estates where there is a superfluity of unsaleable wood, it would be utter folly not to have it converted into firewood, more particularly as such work gives employment to the woodmen when the inclemency of the weather puts a stop to general outdoor work. But this in itself is no proof that the firewood when prepared and ready for the grate is not as expensive as coal; for, when the rent of ground on which the wood was grown, and the cost of felling and converting it into firewood is taken into account, it will be found nearly as costly as household coal of ordinary quality.

What will it cost to prepare a ton of firewood? This is a question that is not readily answered, the cost of labour in various parts of the country varying so widely. In England, generally speaking, the cutting up and stacking of a cord of fairly clean firewood—that is to say when large knotty pieces, which require the mallet and wedge for their manipulation, are excluded—costs from 5*s*. to 6*s*. Again, how many cords of wood will make a ton of firewood? This is another question that is more readily asked than answered, for the difference in weight between equal-sized logs of, say, oak and birch is considerable. For all practical purposes, however, we may state that about one and a half cords of wood go to the ton of firewood, thus making the cost of preparing and housing the latter about 10*s*. The lowest price at which we have sold a ton weight fresh cut was 8*s*., but 10*s*. is nearer the usual price, or about one-half of what is generally obtained for firewood. The cartage of this ton of wood cannot be less than 3*s*. Much depends upon the distance it must be carted, no doubt, but it is usually delivered within a radius of two miles for the price quoted.

The whole matter, therefore, stands something like this: Lowest cost of a ton of wood, 8*s*., cutting same into firewood and stacking, 8*s*.; cost of delivery, 3*s*.—total 19*s*.

It will thus be seen that the difference in price between a ton of firewood and one of coal is inconsiderable, and every one knows which of the two as fuel lasts the longer and imparts the greater amount of heat.

Of course, where the firewood is cut up during wet weather by the estate workmen and consumed on the estate, the matter will stand somewhat differently, the two principal items, the cost of preparing and cutting being considerably diminished. In districts where the firewood cannot readily be sold, and would only rot in the woods, it is a wise policy to have it cut into firewood, not only for the saving effected in the coal bill, but also for preservation of the health of the plantations.

In mining districts, or on the outskirts of large towns, there is usually little difficulty in getting rid of all surplus wood for firing and other purposes, but in thinly-populated, outlying parts of the country, where the cost of transit is excessive, the actual difference between the price of a ton of coal and one of firewood has to be considered.

(2) **Charcoal.**—The following are the chief uses to which charcoal is put in this country: the manufacture of gun and blasting powders, the heating of hall-stoves, cooking, boiling preserves, and the smelting of iron. It is also employed as a filtering and deodorizing agent. Further, it occupies an important place in the making of black paint, ink, ivory- and lamp-black, and is valuable as a horticultural requisite in the packing of bulbs and for potting purposes. In the manufacture of gunpowder, for which a highly inflammable quality is required, the three

principal woods used are the so-called dogwood (*Rhamnus frangula*), the white willow (*Salix alba*), and the common alder (*Alnus glutinosa*), though not infrequently the hazel, chestnut and our native *Rhamnus catharticus* are substituted.

Charcoal produced from the dogwood is, however, preferred to any other, as this forms a very explosive powder, used for military small-arms and for sporting purposes. For this purpose the dogwood is cut when an inch in diameter, and, if possible, when not more than of ten years' growth.

Although iron cylinders or retorts are more economical in the making of charcoal, yet for various reasons the primitive method of pit-burning is to be recommended for general estate purposes. The manufacture of charcoal in this way having received due attention elsewhere in this book, it need not be repeated here. It should be remembered that small wood is more profitable for charcoal-making than that of a larger size, not only because the former requires little or no cutting and splitting, but for the main reason that it can be procured at less cost, and produces more charcoal, weight for weight. Even at the present low price of charcoal—about 1*s*. per bushel—there is a fair profit attached to the making of it, as will be seen from the following figures, which may be taken as fairly representative. A cord of ordinary mixed wood, which should measure, after being stacked, 12 ft. long, 3 ft. wide, and 3 ft. high, will usually, when properly burned, yield 35 bushels of charcoal, and this, at the low price of 10*d*. will realize 29*s*. 2*d*. The cost of cutting this cord of wood, which is generally performed by contract, will be at the least 5*s*., and that of burning 7*s*.; thus leaving a clear profit of 17*s*. 2*d*. per cord for the wood.

Even roots are made into charcoal, and we have seen whole woods grubbed up free of expense, the roots being given in return for the labour; but such work is usually performed during the winter, when labour is at a discount. Pinewood is not nearly so valuable for charcoal-making as hardwood, but the former is not infrequently made to realize a profit of from 8*s*. to 10*s*. per cord of wood.

(3) **Bark.**—In the past, the annual home supply of bark was estimated at about 300,000 tons, but, in addition to this some 30,000 tons were imported from the Continent; but of late years, owing to the employment of chemical substitutes, the amount used is much smaller. For tanning purposes, oak, and occasionally larch, bark is principally in use in this country, though both willow and alder are largely used for the same purpose in various countries, more especially in Russia. Although not at present a valuable product, a small margin of profit, even at the present low price, will accrue through careful and judicious management of the bark crop. Of this we are fully convinced. It is, perhaps, not so well known as it should be that of our two varieties of oak, *Quercus Robur pedunculata* and *Q. R. sessiliflora*—the former contains 15 and the latter only 13 per cent. of tannin. The branches, too, down to an inch in diameter, contain a relatively higher proportion of tannin than the bark of the stem.

The stripping and harvesting of oak bark having received notice in a separate paper, nothing further need be said of these here.

(4) **Faggots.**—These are made of the smaller branches or spray, the remains of charcoal-wood, etc., and tied into bundles similar in size to a sheaf of wheat. They are either left lying on the ground or standing upright in threes or fours together for a few days after being made, previous to being stacked, as they always are for about six months before being used. In thinning a woodland the faggots are usually bound up by contract at 4*s*. 6*d*. per hundred, except when the wood is exceptionally rough and crooked, when another shilling is added. When stacked and dry they realize about 16*s*. per hundred in the wood, thus giving a clear profit of 11*s*. 6*d*., if we deduct 4*s*. 6*d*., for binding, per hundred.

In England the demand for these faggots is considerable, they being used either for kiln purposes, or, when chopped up into smaller bundles, for fire-lighting. These latter are about 9 in. long, and half that in diameter, and are bound tightly round the centre with tarred rope. Previous to the war they were sold at 3*s*. 6*d*. per hundred.

This is a good and profitable way of getting rid of all superfluous spray and branches. Brush or kiln faggots, which are largely used for brick-burning, consist of all refuse woodland scrub, and when tied up and dry can be sold at from 5*s*. to 6*s*. per hundred for the brick-kilns. They are made by contract at 2*s*. 6*d*. per hundred. By the utilizing of this otherwise waste-product, every twig and branch is carefully gathered together, and the woodlands are thus kept in a neat and healthy condition.

Minor Products.—In addition to firewood, charcoal, faggots, etc., which may be considered as the main by-products of the forest and woodland, there are other minor products, such as are to be met with largely where coppice-wood is grown to any extent, which will repay the cost of singling out from amongst the above. These may include flower-stakes, tool-handles, walking-sticks, barrel-hoops, chisel-rods, etc., all of which sell readily in various parts of the country and from which considerable profits are realized.

In cutting the coppice-wood, the longest and straightest poles are selected for hop-stakes, the next size for bean-stakes, pea-boughs, etc., and so on until every part of the wood is utilized.

Leaf-soil, too, sells readily at 5*s*. per cart-load—indeed, near large towns the demand for this and peat often exceeds the supply. In all cases, however, it may not be a wise policy to remove this valuable soil from the woodlands, even at the high price offered.

Half-decayed leaves, too, are much sought after where market gardening is largely carried on, being used to form forcing-beds, and to preserve plants and roots from severe frost.

CHAPTER XXIX

FENCING PLANTATIONS

Many different methods of fencing are adopted throughout the country, each one, no doubt, possessing peculiar advantages according to the circumstances in which it may happen to be required.

In hilly districts very efficient fences of stones may be made where these are abundant. Turf dykes may be constructed on high-lying grounds where stones cannot be readily procured, and iron or wood used wherever fancy dictates.

The term "dead fence" may be applied to these in contradistinction to "live fence" or hedge, to which a special chapter is devoted.

To describe even a few of the various wood or iron fences erected nowadays would be by no means an easy task, but typical examples of several kinds will be explained.

Stone Walls.—These make capital plantation fences, but they are at first rather expensive, and unless well

built require a good deal of attention in the way of repairing breaches. They possess a great advantage over most other fences in the amount of shelter afforded to the young trees. Two methods of building are usually adopted: firstly, where stones are abundant, the entire wall may be of these ; and, secondly, where only a limited quantity are available, the wall is built to a certain height and wires placed atop.

The dry stone wall as this is usually termed, is built without mortar, with the exception of the cope-stone, which in all cases should be bedded in and pointed with lime.

From 4 ft. to 5 ft. is the usual height, the foundations being from 22 in. to 24 in. wide, and the wall 14 in. across beneath the cope-stone, the latter being about 10 in. high and placed on edge. Great care is necessary in building to see that the " throughs " or binding-stones are placed in position, as on this depends mainly the efficiency of the fence. Where wires are used atop, the wall need only be 3 ft. high, 22 in. wide at base, and 12 in. under the cope-stone. The latter are bedded in mortar, and an extra large stone is placed every 6 ft. for receiving the iron standard, to which the wires are attached. Slate slabs, where these are readily procured, may be used for the same purpose as the iron standards, but they should be built firmly into the wall, and reach from the base of the foundation. Two, and sometimes three, wires are used atop of the wall.

Slate Fences.—These are commonly in use throughout Wales ; in fact, wherever slate quarries are worked. When well erected and of fairly regular sized slates, this fence is certainly not to be despised, and it may be considered as practically indestructible. The expenses incurred for keeping these fences in repair are also very little, as they seldom become damaged, and when an upright chances to get broken, another whole one can easily be substituted, and without interfering with any other portion of the fence. The size of slate pale, or slab, as usually termed, is 5 ft. long, 4 in. to 6 in. wide, and about 1 in. in thickness. In erecting the fence a trench is cut about 12 in. wide and 8 in. deep, care being taken that the trench is cut perpendicular, so as to ensure the pales standing in a similar position. These are placed upright in the trench, about 3 in. apart, with their flat side close to the perpendicular cut and the soil replaced in the trench and made firm with a rammer.

A double wire is then tightly interlaced about 3 in. from the top of the pales, and given a double twist between each,

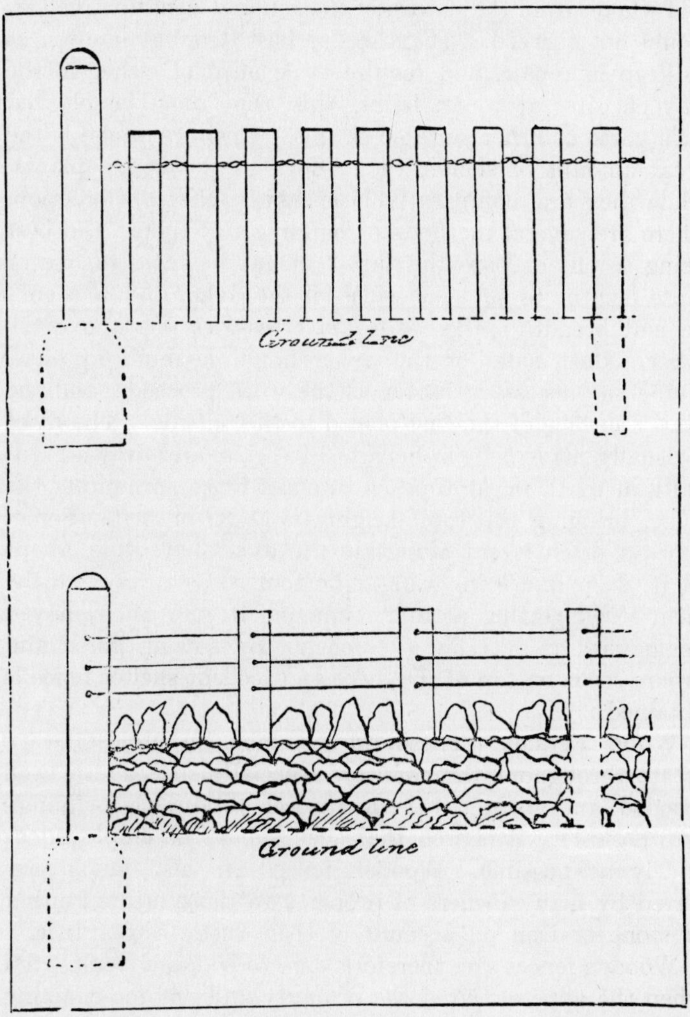

SLATE FENCES

thereby ensuring great stability by uniting the fence and keeping the pales at equal distance apart. The straining-posts are also of slate, 6 ft. long, 6 in. wide, and 3 in. thick.

Turf Dykes.—These were formerly much used in moorland and outlying districts, where stones are not abundant, and where, from the nature of the soil and situation, hedges would not succeed. They are, at best, troublesome fences to keep in repair, and require some adjunct either in the way of wires atop, or, failing this, they must be planted with gorse or other suitable shrubs. One advantage is the great amount of shelter they afford to the young plants, while they are, comparatively speaking, cheap of erection. There are several methods of building turf dykes, the best being to cut or pare the turf 3 in. in thickness in one or more lengths to suit the width of the dyke, and of a convenient breadth: these are laid cross-wise one above the other. Both sides of the dyke should be built at once, giving the necessary batter as the work proceeds, and the grassy surface of the turf placed to the outside. The dyke is usually made 3 ft. in height, 3 ft. wide, and drawn gradually in to 12 in. at top. A two-rail fence surmounts the dyke, bringing the total height to 4½ ft. or 5 ft. Sometimes a ditch is cut alongside the dyke 3 ft. wide, about 2½ ft. deep, and 9 in. wide at bottom, so as to prevent the farm stock getting at and damaging it, the soil removed being used in forming the fence. By sowing gorse and broom seeds on top of the dyke an excellent shelter fence is obtained.

Wood Fences.—These are common on almost every estate throughout the country, especially such as are well wooded, and, owing to the low prices obtainable for home-grown timber, it is well that such should be employed as widely as possible. Wooden fences are also much preferred by many owners of property to those erected either of stone or iron on account of their rustic appearance.

Wooden fences are, therefore, sure to be largely employed when the appearance of the property and not too-exacting financial results are points of importance.

There are many forms of wooden fences adopted, these varying chiefly according to the particular use to which they are applied. The following descriptions are of such kinds as are generally in use for woods and plantations.

Here it might be well to mention in passing that only matured and seasoned timber should be used in fencing, the cost of erection, whether the timber be good or inferior, being the same, and every one knows which will last the longer.

A good strong fence is erected as follows :—Posts, 5 ft. 9 in. long, 4 in. broad, and 2½ in. thick ; bars or rails, 9 ft. long, by 3½ in. by 1½ in. Four holes are mortised into the posts for the reception of the bars, the ends of which are so formed as to overlap each other tightly. The fence is usually 4 ft. high, and so as to strengthen the horizontal

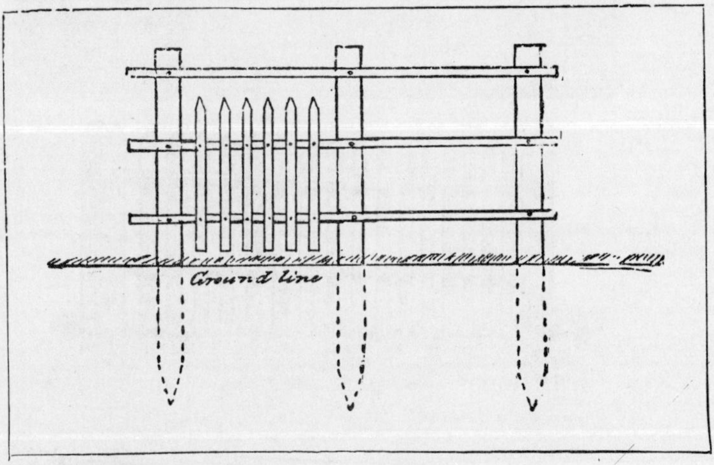

SAWN WOOD FENCE

bars a stake is driven into the ground midway between the larger posts, and to this the rails are securely nailed. In some cases the posts are not to be mortised, so that the bars require to be attached by nails.

For park clumps, particularly where a substantial and neat fence to keep back horses, cattle or deer is required, the following, though rather expensive at first, is largely employed. The entire fence is made of oak or Spanish chestnut, and is shown on following page.

Posts 7 ft. long, 6 in. by 4 in., and run out with the circular saw. Rails triangular, about 3½ in. to the side.

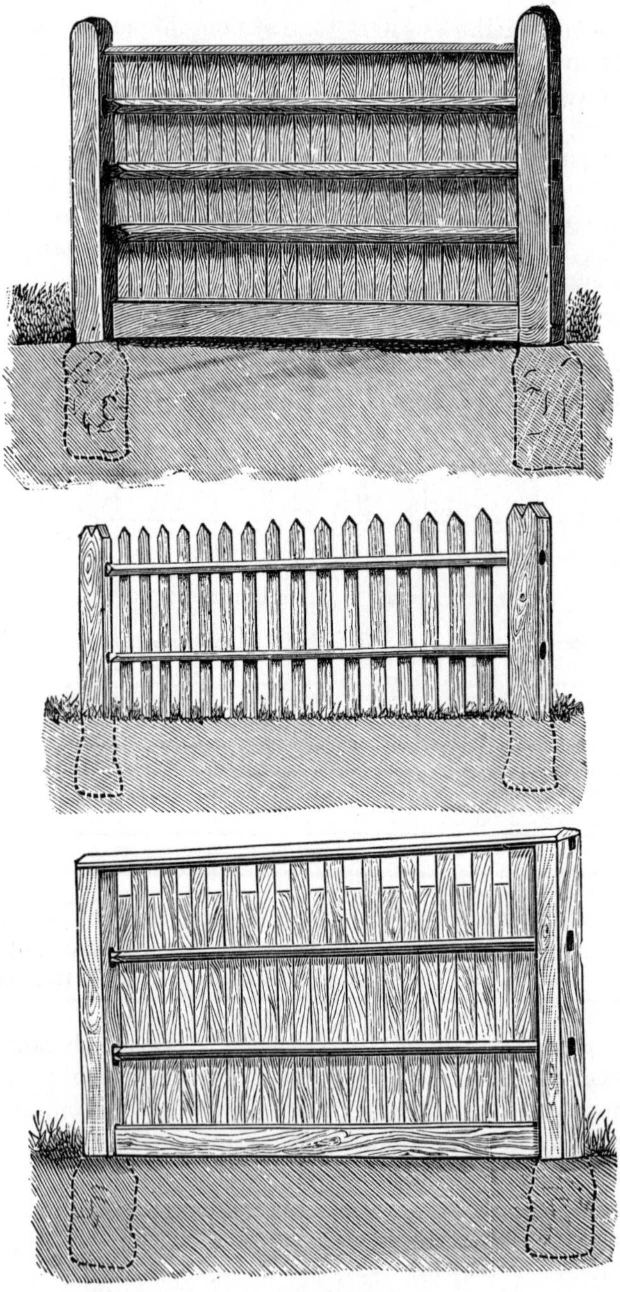

OAK FENCES

The uprights are rent from oak or chestnut trees of straight grain, and are usually about $\frac{3}{16}$ of an inch thick, and 5 ft. high. The posts are erected 6 ft. apart, the rails being mortised into these, and the rent uprights fastened about 2 in. apart by patent rose nails to the horizontal rails. A fence of this kind, when properly erected, will last for upwards of forty years, especially if the butts of the posts are charred before being inserted in the ground.

Rustic fences for small tree clumps may be of almost

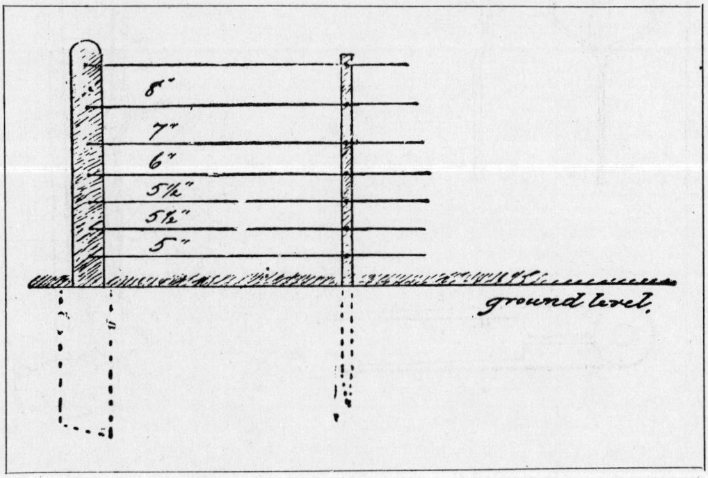

WIRE FENCE WITH WOODEN POSTS

any design, but the following is cheap and easily erected. It is formed of larch posts 6 ft. long, and about 4½ in. diameter, driven into the ground at 6 ft. apart. Two flat or rounded rails about 3 in. by 1¼ in. are nailed horizontally to these, the lower at 9 in. from the ground and the other flush with the tops of the posts, which when driven in are 4 ft. from ground level. The uprights are also of larch, split up the centre and nailed on the horizontal bars at 2 in. apart. They extend above the top rail for 7 in., and are sharply pointed so that they cannot be climbed over.

Wire Fences.—These may be erected either with iron

or wooden standards and straining-posts. The form most commonly in use for enclosing woods is that with wooden posts and strainers, these being made of mature and thor-

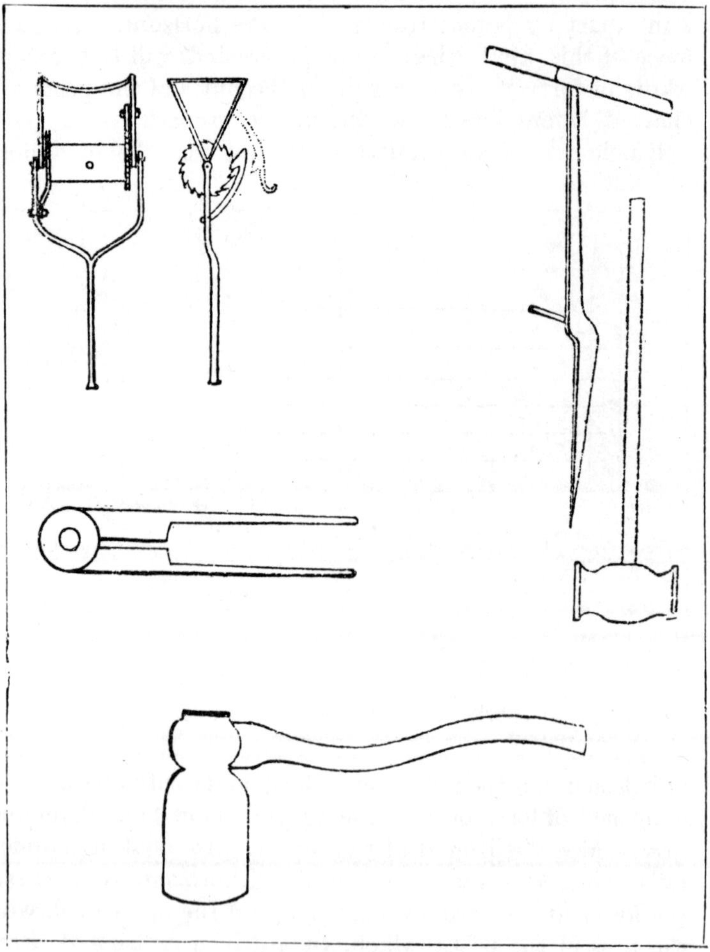

TOOLS FOR FENCING

oughly seasoned larch or oak. The strainers are 7 ft. long and 6 in. square, or, if round, about 7 in. in diameter, while the intermediate posts are $5\frac{1}{2}$ ft. long, and $3\frac{1}{2}$ in. by 3 in., or, if round, $3\frac{1}{2}$ in. diameter at smallest end. The strainers should be fitted into the ground at 150 yards apart, and the posts driven firmly at 6 ft. from each other. At every sharp curve along the line of fence a stout post, say 5 in. in diameter, should be used. In order to make a stout fence proof against cattle and sheep, six wires should be used, the two top No. 6, and the others No. 7 gauge, the distances between each pair, beginning at the top, being 8, 7, 6, $5\frac{1}{2}$ and 5 in., the lower being 5 in. from the ground. Brackets for straining the wires should be attached to each of the strainers, these having this advantage over the older system of using the straining machine, that the wires can be loosened or tightened at will, when repairs are found necessary. The tops of the posts should be rounded off or sawn on angle so as to prevent the lodgment of water. Iron and wire fences combined are now commonly in use, and there are so many excellent systems that it would be invidious to recommend one kind more than another.

Iron box fencing, which consists of standards with double pronged feet for fixing in the ground and round or flat horizontal bars run through them, has been largely used and looks neat, being also, if properly erected, very efficient. Wrought-iron hurdles are sometimes used for fencing park clumps, and they possess at least this advantage, that they can be lifted at any time and re-erected should it be found necessary to remove them from one place to another.

Unclimbable iron fencing, usually in hurdles 7 ft. long, are now much in use for park fencing, but for general plantation purposes this class of fencing is too expensive.

Tree Guards.—These may either be erected of wood or iron ; the former is, however, preferred on most large estates where timber is plentiful, and will receive first attention. For large trees whose branches sweep the greensward an elaborate structure is required, which may take the form of almost any of those described under wooden fencing. That entirely formed of oak is to be recommended, or split larch for uprights, with oak posts and rails may be considered more rustic in appearance. In any case the guard should be sufficiently high and wide to prevent cattle and horses reaching over to damage the branches.

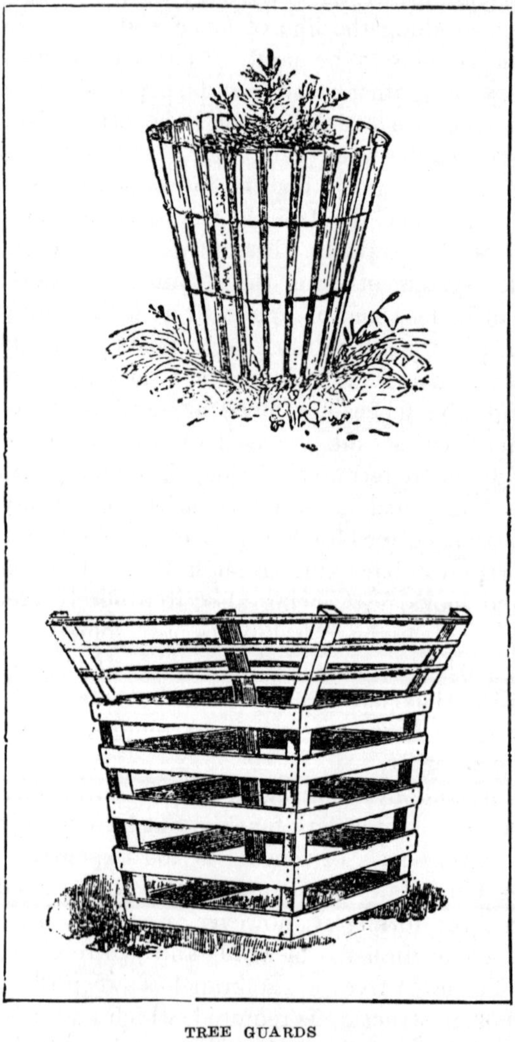

TREE GUARDS

When the trees are destitute of branches for a considerable distance up the stem, say 8 ft. or 10 ft., a very neat and efficient guard is made as follows :—Procure a number of larch, oak or Spanish chestnut poles, 7 ft. high, and about $2\frac{1}{2}$ in. diameter at small end. Thread these on wires by boring holes in the poles at 2 ft. and 5 ft. from the butt end, keeping each couple separate by 3 in.-long pieces of the same size of pole, also threaded on the wires. These can be formed on the level, and when sufficient to embrace the tree have been got together, the whole may be lifted up and placed in position closely around the trunk. Another method is to bind the poles together with fencing wire, giving a double twist between each to keep them at a suitable distance apart.

When a more elaborate fence or guard is required, four posts 7 ft. long, 4 in. square, and sawn from crooked oak branches are used. The posts are quite straight for 5 ft. in length, the upper 2 ft. being inclined outwards, which not only gives the guard a neat appearance, but is a great preventive against the encroachments of farm stock. The posts are inserted nearly 2 ft. in the ground, and so as to form a square around the stem of $4\frac{1}{2}$ ft. to the side, four bars, each 3 in. by 1 in., are nailed horizontally on the straight portions of the posts, and at equal distances apart from where the angle occurs downwards ; upwards from that three hoop-iron rails are nailed in a similar manner, the top one being one inch below the level of the crown of the posts. Iron has a light and neat appearance when used for the top bars, but wood is often substituted. Another cheap and neat tree guard for using with the rarer trees, to which horses and cattle have not access, is made as follows :—Pales 3 ft. long, 2 in. wide, by $\frac{3}{8}$ in. thick, are sawn out and pointed. They are driven into the ground round the tree to be protected, the tops sloping outwards and $1\frac{1}{2}$ in. apart. Stout tying wire is then interlaced at two heights from the ground.

Oak or chestnut bark placed loosely around clean-stemmed young trees will prevent damage by ground game, and is cheap and looks unobtrusive.

CHAPTER XXX

TIMBER MEASURING

To those who are not practically acquainted with the measuring of home-grown timber the following brief remarks in elucidation of the subject will be useful. It may, however, be well to mention that timber measuring is rather a vexed question, some following what is known as Hoppus's system, and others advocating that of Horton. The former being that generally in use amongst timber merchants in this country, and consequently of greatest value to the forester, the following details of this system may prove serviceable.

Regarding the timber-measurer's equipment it may first be necessary to say a few words. This consists of a 66 ft. Chesterman's tape-line, or instead of this a 5 ft. wooden rod, standard girt-strap, or fine cord, scribing knife, and bent piece of iron, with eye at end for drawing the girt-strap beneath such trees as the arm cannot readily pass under.

For girthing timber a piece of thin whipcord or string is frequently used, but as the elasticity of this varies greatly, and has in many instances led to dispute, a much fairer plan and one that is liable to no abuse is to use the 12 ft. girt-strap, upon which every inch in length is reckoned $\frac{1}{4}$. As the proper quarter-girth can be seen at a glance on this strap, its adoption will at once remove any chance of trickery, which may be possible in the use of the string and rule.

For entering measurements the most convenient book

is that 9 in. by 4 in., with stiff pasteboard covers, ruled with horizontal lines, and divided into four vertical columns. To measure proceed as follows :—Mark a number with the scribe on the butt end of the tree, and enter a similar number in the first column of the book ; this will not only serve to identify the particular log, but prevent any risk of measuring twice. Should the taper throughout the whole length of the tree be tolerably gradual, set down the length in the second column of the book, opposite the number already entered. At exactly one-half of the length of the portion measured, take the girth by passing the girt-strap tightly around the stem. Put this down in the third column on the same horizontal line as the number and length. Should, however, the trunk taper not be fairly uniform throughout the entire length, as frequently happens, several measurements may require to be taken.

For example, a tree may be 36 ft. in length, running with regular taper for perhaps 12 ft., after which it branches out, reducing the size of the remaining part very considerably for, say another 12 ft., where it again branches and leaves the last 12 ft. of a relatively small size. With such a tree it would be quite impossible to obtain anything like a correct measurement by taking only one length and girth. The difficulty is, however, readily got over by first measuring the lower 12 ft., then the second, and then the third, marking the respective lengths and girths in the vertical columns as already described.

The measurement of these trees, so far as the field work is concerned, is now completed, the contents of each tree being found by referring to "Hoppus's Measurer"—a book with which every forester should be supplied. By squaring the quarter-girth in inches, multiplying by the length in feet, and dividing by 144, the same result will be obtained. This is, however, a tedious method, especially where large numbers of trees have to be dealt with, and should only be adopted when Hoppus is not at hand.

By committing to memory the following short table of quarter-girths much time in calculating and consulting authorities will be saved.

Quarter-girth			Contents equal to … the entire length in feet
6-inch quarter-girth	will give contents	equal to	$\frac{1}{4}$ the entire length in feet.
7 ,,	,,	,,	$\frac{1}{3}$
$8\frac{1}{2}$,,	,,	,,	$\frac{1}{2}$
10 ,,	,,	,,	$\frac{2}{3}$
12 ,,	,,	,,	1
13 ,,	,,	,,	$1\frac{1}{4}$
$14\frac{3}{4}$,,	,,	,,	$1\frac{1}{2}$
16 ,,	,,	,,	$1\frac{3}{4}$
17 ,,	,,	,,	2
19 ,,	,,	,,	$2\frac{1}{2}$
21 ,,	,,	,,	3
$22\frac{1}{2}$,,	,,	,,	$3\frac{1}{2}$
24 ,,	,,	,,	4
27 ,,	,,	,,	5
$29\frac{1}{2}$,,	,,	,,	6
$31\frac{3}{4}$,,	,,	,,	7
34 ,,	,,	,,	8
36 ,,	,,	,,	9

The sliding rule is also useful for determining contents.

The proper allowance to be made for bark is half an inch for every foot of quarter-girth for oak and elm under 12 in. quarter-girth, and an inch for all beyond, but it is quite impossible to fix upon any one uniform scale that will meet even the majority of circumstances. Trees growing in exposed situations will frequently have bark almost double the thickness of those of a similar size in the woodland. My plan has been always to allow for the bark of each tree at the time of measurement.

Measuring Standing Timber.—For this a pliable pole 18 ft. long, marked in feet, and the girt-strap already referred to are the necessary equipment. In estimating the number of feet of timber upon a large area, it is not always necessary to measure each tree separately, particularly when the whole situation is composed of one species, and the individual trees are about the same age and size, as by multiplying the total number of trees by the average content of those selected and measured a very just calculation will be arrived at. Great care in their selection, and considerable judgment in taking the average will, however, be required.

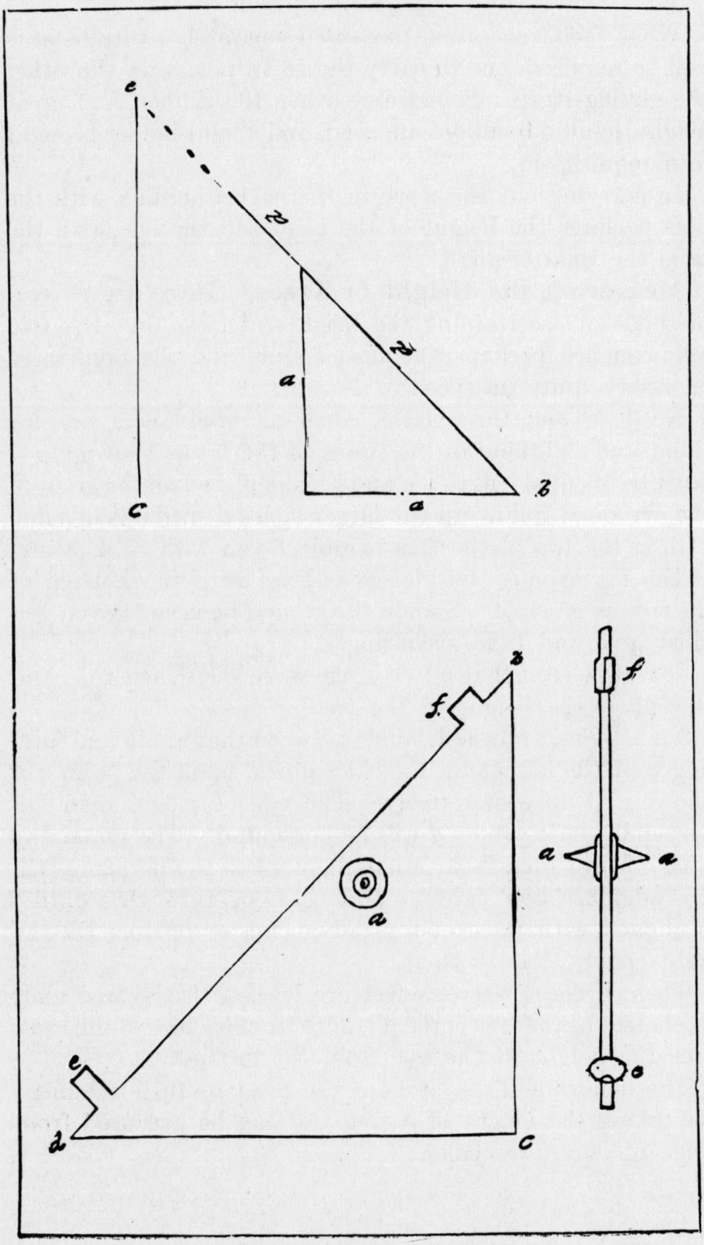

MEASURING HEIGHT OF TREES

When each tree is to be measured separately two assistants will be required, one to carry the 18-ft. pole, and the other the girting-strap. Sometimes, when the timber is of great height, jointed bamboos are used, and a light ladder brought into requisition.

In carrying out the work in this way, the man with the pole declares the height of the tree, and the one with the tape the quarter-girth.

Measuring the Height of Trees.—There are several methods of ascertaining the heights of trees, but the two following are, perhaps, the most simple, and the appliances necessary quite inexpensive :—

No 1.—Take three laths, such as bricklayers use for tiling, and nail them in the shape of the frame shown ; *a a* must be of equal length ; *a* and *b* being placed on the ground, the eye must follow up the larger lath *d d* until it is in a line with *e*, the top of the tree or object you wish to measure.

The frame must be placed as level with the bottom of the tree as possible. Should the ground be very uneven you must give and take accordingly.

It will be seen that *b* to *c* is the same length as *c* to *e*, and this gives the height of the tree.

No 2.—Suspend the triangle between the thumb and fore-finger of the left hand, knuckles down, upon the point *a a*, allowing it to swing freely. The edge *b c* will then fall perpendicularly, and *c d* will be horizontal. The remaining edge *d b* will then lie at an angle of 45 to the horizon. On this edge are two sights, *e* and *f*. Look through *e* until *f* is aligned with the tree-top, advancing or retiring till the sights point exactly to it.

Then, if the observer's feet are level with the tree root, the height of the tree is the distance from his feet to the root, plus the height of the eye from the ground.

The dendrometer is perhaps the most useful instrument for taking the height of a tree and can be procured from some of our nurserymen.

CHAPTER XXXI

BLASTING AND BURNING TREE ROOTS

BLASTING by gunpowder or dynamite is not only the most expeditious but also the cheapest method of clearing away tree stumps and large logs. In preparing to blast a stump, great care must be exercised to bore the hole in the right place and not to use too much explosive. For blasting powder the hole should be $1\frac{1}{2}$ in. in diameter, and should penetrate to the centre of the stump. It must not be too low down, lest the bottom should blow out and the force be expended in shattering the ground instead of the stump or log. In selecting the spot to bore for the powder, choose the hardest part of the root and ensure an equal thickness

BLASTING AND BURNING TREE ROOTS

of wood all round, and even splitting of the log will be the result. The following is a good way of putting in the powder :—For large stumps of from 2 ft. to 4 ft. in diameter about $3\frac{1}{2}$ in. depth of coarse blasting powder should be inserted in a hole $1\frac{1}{2}$ in. in diameter. The end of the fuse should be put into the centre of the powder, and left protruding for 15 in. outside the hole, which is filled with dry sand, consolidated, or packed around the fuse by means of a coarse iron wire. The outside end of the fuse should be teased out and lighted with a match, and as it will require over a minute for the fire to reach the powder, time is given for the operator to find a place of safety.

Burning Tree Stumps.—With a 2-in. auger bore a vertical hole in the centre of the stump from the top towards the bottom. In the side of the stump, near ground-level, bore a horizontal hole towards the centre, so as to open into the vertical hole, drop some fire down the vertical hole, and if the wood is at all dry the draught of air entering by the horizontal hole will, like the draught of a chimney, maintain the combustion of the fire in the centre, until this slowly spreads and ultimately burns away the stump.

Another and equally simple method of destroying stumps of trees is as follows :—In autumn bore a hole 2 in. in diameter and 18 in. deep, put in 1½ oz. of saltpetre, fill with water, and plug up close. In the following spring put in the same hole half a gill of kerosene oil and then light. The stump will smoulder away without blazing, down to every part of the roots.

American Method of Blasting.—At Studley Horticultural College, Warwickshire, the American method of blasting was successfully carried out and reported upon by Mr. A. P. Long as follows :—

A hole is bored with a long auger or crowbar in a sloping direction from one side of the stump to its base, generally from 2½ ft. to 3½ ft. deep. The bore-hole is cleaned out, and a number of dynamite cartridges inserted, each being firmly pressed home by a wooden rod. A primer cartridge containing a detonator is then placed on the top of these, and the bore-hole is filled with clay and tightly rammed. The primer is either connected directly with a safety fuse, or to a high-tension battery, by a cable, and is afterwards fired. As dynamite strikes downwards as well as upwards, the effect of the explosion is that the roots and stump are all either ejected or loosened, so that they can be easily removed by hand.

The American method is less costly and more speedy than the methods hitherto used in England in removing stumps. If there is no man on the estate qualified to handle explosives, an expert must be employed at about £1 per day, besides travelling and hotel expenses. Three men —an expert and two labourers—can bore holes and blast thirty sound stumps per day easily. If the stumps are hollow in the centre, two or three bore-holes are necessary for each stump, and in that case twenty only can be blasted during the day. Taking the pre-war wages of two labourers at 2*s.* 6*d.* each per day, the cost of boring and firing averages 2½*d.* per stump, exclusive of the expert's fee. The expert's fee increases the cost by about 2*s.* per stump.

The explosive used is Nobel's dynamite, in the form of cartridges, costing 9½*d.* per lb. The average quantity used for each stump is between 2 lbs. and 3 lbs. (about twenty to thirty cartridges), so that the cost of the explosive is not more than 2*s.* 6*d.* per stump. The detonators and fuses required only cost a few pence. Summing up, the cost per stump is :—

	s.	*d.*
Expert's fee	2	0
Cost of boring	0	2½
Cost of explosive	2	6
Detonators and fuse	0	9½
	5	6

Misfires and partial removal of stump may require fresh borings and further charges of explosive, thus increasing the cost. By employing a skilled estate hand capable of using explosives instead of an expert, the expense, however, is greatly diminished.

By the old method of grubbing and jacking, stumps were removed at Studley some time ago at the high cost of about £2 5*s.* each butt, and even then success was only partial. In another case, on an estate in Norfolk, where an old pasture was converted into a plantation of mixed trees, trenching at the cost of £18 per acre had to be resorted to on account of the presence of roots and stumps of old trees. In this case it would have been much cheaper to have removed the stumps by blasting. The demonstrations at Studley showed that both sound and unsound stumps could be successfully blasted, and whole trees—an Apple and an Oak—were also uprooted by the same method with equal success, using only one bore-hole and about the same charge of explosive. The timber of the trees so treated, however, is very much split, so that blasting is only advisable when the timber is considered of little value.

The particular explosives used are unaffected by damp, and, in consequence, the method is applicable in both wet and dry situations. Firing the charges was done at the demonstrations mostly by ladies, and a photographer was able to get sufficiently near to obtain photographs of the effect of the explosion without danger. The principal recommendations of this method, therefore, are cheapness, effectiveness and safety.

CHAPTER XXXII

PRICES OF CONTRACT OR PIECEWORK

The following prices may be taken as approximating to those paid generally throughout the country previous to the war.

It may be well to remember, however, that in districts where unusually high or low wages are paid, so in proportion will be the contract prices for the various classes of work.

BARKING OAK:		s. d.		s. d.
Barking oak	per ton	21 0		
Loading bark on wagons	,,	3 0		
Barking oak per ton of bark		30 0		
Chopping bark	,,	8 0	to	10 0

COPPICEWOOD:		
Cutting out hurdle rods	per score	0 1
,, hurdle stakes	per dozen	0 1
,, rake stems	,,	0 1
,, spade stems	,,	0 1
,, spick gads	,,	0 1
,, dahlia stakes	,,	0 1
,, rose stakes	per two dozen	0 1
,, besom handles	,,	0 1
,, kidney-bean stakes	50 in bundle	0 2
,, pea stakes	per bundle	0 1
,, birchwood for besoms	,,	0 1

DRAINING:				
Pipe draining, mains, 4 ft. 3 in. deep, 4 in. or 6 in. pipes	per chain	3 0	to	3 6
,, ,, small, 4 ft. deep, 1½ in. to 3 in. pipes	,,	2 0	,,	2 9
,, ,, small, 4 ft. deep, 1½ in. pipes	,,	1 9	,,	2 6

		s. d.		s. d.
Ditches, open, 36 in. wide at top, 30 in. deep, and 9 in. wide at bottom	per chain	3 0	to	4 0
Scouring out ditto . . .	,,	0 9	,,	1 0
Small open ditches, 15 in. to 18 in. wide at top, 12 in. to 15 in. deep, and 9 in. wide at bottom. .	,,	1 6	,,	2 6
Scouring out ditto . . .	,,	0 6	,,	0 9

FAGGOT-MAKING :

Making faggots	per 100	4 6
,, oven faggots	,,	3 0
,, faggots for fire-lighting . . .	,,	1 3
Cutting bands for tying faggots . . .	,,	0 4

FELLING AND STUBBING TIMBER :

Felling oak timber .	per ton of 40 ft.		3 0		
,, other hardwoods	,,	,,	2 6		
,, pinewood .	,,	,,	1 6	to	1 9
Stubbing out timber .	,,	,,	3 6	,,	4 0
Cutting underwood from 12 to 15 years' growth	per acre		9 6	,,	12 0

FENCING :

Setting out and mortising 4-holed posts	per score	4 0		
,, ,, ,, 3-holed ,,	,,	3 0		
,, ,, ,, 2-holed ,,	,,	2 6		
,, ,, ,, 1-holed ,,	,,	2 0		
,, and cleaving rails . .	,,	0 10		
,, ,, ,, stakes .	,,	0 6		
,, ,, ,, long poles	,,	0 6		
,, ,, ,, short poles	,,	0 4		
Hanging field-gates	each	4 0	to	5 0
Fixing stile	,,	2 0		
Preparing posts, rails and pails for tree-guards	per set	0 9		
Fixing ditto	,,	2 0		
Six-wire fence larch posts and creosoted	per chain	35 0	,,	40 0
Fixing same	,,	4 6	,,	5 0

		s.	d.		s.	d.
FIREWOOD:						
Splitting firewood	per cord	4	0	to	6	0
,, ,, for charcoal .	,,	2	6	,,	3	0
Cutting and stacking cordwood .	,,	2	0	,,	3	0
Burning charcoal .	per bushel of 20 lbs.				0	2
Loading and spreading soil . .	per load	0	5	,,	0	6
GATE-MAKING:						
Field gate, oak	5-bar				17	0
Posts	per pair				23	0
Iron fastening and ironwork complete . .					5	9
,, Fixing.					4	6
Making 5-bar oak gate.					2	6
,, half gate.					1	9
,, rough wickets					2	0
,, wrought wickets . . .					3	0
Sawing hardwood. . . .	per 100 ft.	3	6	to	4	0
,, softwood	,, ,,	2	6	,,	3	0
HEDGING:						
Trimming hedges, ordinary size .	per chain	0	9	to	1	6
Making bank for quick hedge, digging ditch and planting quick	,,	7	6	,,	10	6
,, hedge, without bank or ditch, trenching ground, preparing bed and planting	,,	3	0	,,	4	0
Cleaning young hedges. . .	,,	0	8	,,	0	9
Laying hedge and scouring out ditch	,,	2	0	,,	3	0
HURDLE-MAKING:						
Making hurdles	per dozen	3	6	to	4	0
,, cattle hurdles (wattle) .	,,	5	0	,,	6	0
,, Welsh hurdles, for sheep .	,,	4	0	,,	5	0
PITTING:						
Digging out clay	per yard				0	6
,, holes for tree-planting, 15 in. in diameter and 15 in. deep . .	per 100	1	6	to	2	6
Inserting plants	,,	0	6	,,	1	0
Notch planting	,,	2	0	,,	3	0
PREPARING ROAD MATERIAL:						
Quarrying stones	per yard	0	10	to	1	3
Breaking stones for roads . .	,,	0	8	,,	1	0

CHAPTER XXXIII

FOREST AREA OF THE WORLD

EXCLUSIVE of the forests of China, Corea and parts of Africa and South America, for which there are no available data, the forest area of the world is approximately 3,800,000,000 acres.

The forests of Europe total upwards of 750,000,000 or, roughly speaking, about 31 per cent. of the total land area of the Continent; whilst among non-European countries Canada comes first with 799,000,000 acres, United States, 545,000,000, tropical South America 528,000,000, Asiatic Russia 348,000,000, and Central Africa 224,000,000.

Finland is, perhaps, the best wooded country in the world, Bosnia, Herzegovina and Sweden coming next, whilst amongst the least wooded areas are Great Britain and Portugal, the former including only about 4 per cent. of the total area of the land.

So far as is at present known the following are the approximate areas of woodlands in the various countries of the world :—

COUNTRY.	TOTAL FOREST AREA. Acres.
Russia—	
European Russia	461,611,000
Finland	52,500,000
Austria-Hungary—	
Austria	23,996,000
Hungary	18,692,000
Croatia and Slavonia	3,769,000
Bosnia and Herzegovina	6,380,000
Sweden	49,390,000
Germany	34,990,000
France	24,021,000
Norway	16,848,000

Country.	Total Forest Area.
	Acres.
Spain	16,065,000
Italy	10,115,000
Bulgaria	7,603,000
Roumania	6,367,000
British Isles	3,071,361
Switzerland	2,140,000
Belgium	1,304,000
Servia	3,865,000
Other Countries	4,427,000
	747,154,361
Asiatic Russia	348,030,000
India	149,000,000
Ceylon	6,763,000
Japan	57,718,000
Philippine Islands	49,000,000
British Australasia	126,720,000
Cape Colony, Natal, Swaziland and Transvaal	641,000
Madagascar	25,000,000
Barbary States	9,527,000
Central Africa	224,000,000
South America (tropical)	528,000,000
West Indies	42,669,000
Canada	799,360,000
Mexico	25,000,000
Alaska	107,000,000
United States of America	545,000,000
Other Countries including the Straits Settlement, Java, etc.	6,870,000
	3,050,298,000

The approximate area of woodlands in Great Britain and Ireland are :—

	Acres.
England	1,715,473
Scotland	868,409
Wales	184,361
Ireland	303,118
	3,071,361

CHAPTER XXXIV

AFFORESTING WASTE LANDS AND THE FINANCIAL RETURNS THEREFROM

Now that the Government is being urged seriously to consider the question of afforestation, it may be opportune on my part, as one of the earliest writers on the subject, to briefly recall what has already been done in this matter, and to offer some remarks on planting waste lands, with special reference to cost and the financial returns therefrom.

For the past thirty years I have not failed to urge on the State and private owners of woodlands the pressing necessity for planting up some at least of the waste and unprofitable lands of our country, in order to provide a sufficiency of timber for the future and leave us less dependent on the supplies that are annually sent us from abroad. As stated elsewhere, when we consider that the total area of woodlands in this country is only a little over 3,029,000 acres, that fully 15,000,000 acres of waste lands exist, and that we annually import over 10,000,000 tons of timber, at a cost of about £25,000,000, the necessity for an increased area of woodlands, so that a portion at least of this vast sum may be kept at home, will be apparent to all, and the more so as a dearth of timber is imminent, and outside supplies are being rigidly conserved, while our home demands are ever on the increase. England being, so to speak, a residential country, the retention of a certain amount of heath, mountain and common lands, for the purpose of deer forests, grouse moors, game coverts and golfing links is imperative, and will considerably reduce the acreage of land available for afforesting purposes. But I think that I am well within bounds in alloting out of the

15,000,000 acres of waste land 1,000,000 to afforesting and 14,000,000 to game preserves, deer forests and rough pasture.

Having personally explored much of the mountain and heath lands in England and Scotland, and some of the vast tracts of bog land in Ireland (the latter extending to fully one million acres), I have carefully computed that of land up to 1,200 feet altitude, where timber would grow perfectly well, about 9,000,000 acres are available for afforesting purposes. As far as I have been able to find out, the average rental of the ground referred to is a fraction under 3*s*. per acre, and I am quite confident that any land which does not bring in at least three times that amount for grazing or agricultural purposes would be more profitably employed in carrying a crop of timber.

It is unfortunate that much of these waste lands are private property, the owners of which, even could they afford it, have little inclination to sink, for a period of say twenty years, the necessary capital required to be expended on the formation of woods and plantations. Equally unfortunate is it that owing to an injudicious system of management many plantations in this country have been wrongly formed—in so far as adaptation of soil and trees are concerned, the results being that financially speaking the woods are a failure, and proprietors in consequence fight shy of further planting operations. I have examined and reported on several of such woods in various parts of the country, one of the most noticeable being in Nottinghamshire, where a large area of ground was planted with a crop of oak, for which tree the soil was quite unsuitable, the result being that over the whole ground the average production of timber per tree was under 10 cubic feet in sixty years. When pressing home the question of woodland extension I have frequently been confronted by the argument that past experience does not warrant further expenditure in that way. That this is true cannot be denied, but let us hope that it will be remedied in the near future by the better education of our foresters and by greater attention being given to the relation of trees and soil.

With the wholesale felling of timber for war purposes

and the disinclination of owners of land to engage in extensive planting operations, the question naturally arises: What is the most feasible way of overcoming the difficulty?

In answer, and without the slightest hesitation, I would say that the State should acquire and plant suitable waste lands at the rate of 40,000 acres annually for a period of twenty-five years. Such lands could, in England, Scotland and Wales, be gradually and cheaply acquired by the State, while in Ireland there are vast tracts of peat bog which their owners would willingly hand over to the Government at the present time at a small cost per acre. Taking the British Isles as a whole, the cost of procuring suitable lands would be at an annual rental of about 3*s.* per acre, or 40*s.* per acre for purchase. On the Gwydyr Estate, Carnarvonshire, 7,412 acres of land, described as rough grazing and sheep walk, were lately sold by public auction for £15,670, or at the rate of £2 2*s.* 3*d.* per acre. I have little faith either in the State advancing money to landed proprietors towards afforesting, or in municipalities coming to the front as planters of woodlands. The State would be the best custodian of forest property for the simple reason that the State only can readily acquire the needed land in sufficient quantity and on the best terms, and I am fully convinced that plantations formed under Government supervision will, in an economic sense at least, be far more successful than those planted either by private persons or public bodies. Again, the continuity of ownership under such a scheme, together with the ample resources guaranteed by State control, would both largely contribute towards a successful issue in such an undertaking.

The difficulty of housing and providing for the workmen employed in afforesting out-of-the-way lands has been brought to my notice, but from personal experience of similar work in Scotland and Wales I anticipate little difficulty in that way. In these cases, where a good deal of the work was carried out by contract, the workmen gladly walked to and from the adjoining villages each day, often a distance of three or four miles, bringing their midday meal with them, which was heated or cooked on the ground. Then, as the plantations increase in age and size, and sawmills are required, the ever-increasing industry so created will cause hamlets to spring up in the wooded regions, just as we find is the case in mining and quarrying districts.

After careful computation I have no hesitation in saying that the area of plantations in the United Kingdom could at once be doubled by the planting of waste lands which at present do not bring in over 2*s*. per acre per year of rental, with infinite benefit to the country generally and a vast increase in the value of land both to the owner and farmer who cultivates it. In the matter of afforesting, a grain of practice is worth a ton of theory, and as I have personally supervised every operation, from marking out the plantation boundary on the exposed hillside, to draining, fencing, planting, thinning and disposing of the produce, my opinions on the question are at least worthy of consideration.

The Approaching Scarcity of Timber.—Than timber no article is probably more indispensable to the welfare of a nation, entering extensively as it does into almost every trade and industry. For England, therefore, with an ever-increasing import, the possibility of a dearth of timber must be regarded with the keenest anxiety, more particularly as this would entail prohibitive prices and seriously cripple the trade of the country. The following table, as reported to the Washington Bureau of Manufactures, will show at a glance the annual imports of timber of the principal countries of Europe :—

England .	16,342,600	cub. yds.	Italy . . .	915,148	cub. yds.
Germany .	11,766,667	,,	Denmark .	849,630	,,
France .	8,496,300	,,	Spain . .	392,222	,,
Belgium .	1,897,777	,,	Switzerland .	313,778	,,

In face of this it is only reasonable to suppose that the Government will act promptly in the matter, remembering that no scheme of afforesting, however extensive or well ordered, can bring the necessary relief for at least forty years after its inception. For all this, and in spite of numerous warnings as to the pressing necessity for tree planting and the ominous signs of a timber famine, little or nothing has been done, save the holding of meetings by the Board of Agriculture and the purchase of a few hundred acres of waste land in Scotland.

To sum up briefly, the situation is this :—England's imports before the war rapidly increased from a trifle under $3\frac{1}{4}$ million loads in 1864 to fully 10 million loads in 1906, thus showing an increment of fully 7 million loads in forty-two years.

Most European countries have large internal supplies of timber, so that, by a system of conserving and protective tariffs, the pinch of want would not be felt severely for years to come. But not so England, which is almost wholly dependent on supplies from abroad.

According to the Secretary of the Agricultural Department of Washington, the area of forests in the United States is 700 million acres, but even now the States are more or less dependent on Canada, and actually receive the entire surplus from that country. But regarding the United States, ex-President Roosevelt said : " If the present rate of forest destruction is allowed to continue with nothing to offset it, a timber famine in the future is inevitable. Remember that you can prevent such a famine occurring by wise action taken in time ; but once the famine occurs there is no possible way of hurrying the growth of trees necessary to relieve it." Again, the late Mr. Lewis Miller, who had vast forests both in Sweden and Nova Scotia, told me that in twenty-five years neither the United States nor Canada will have much timber left, while Sweden and Finland are already played out. "I am also of opinion," he said, " that during the next twenty-five years timber will be double its present price, and that it will not only pay to plant land valued at 3*s*. per acre, but that worth 20*s*. per acre." These are no idle words, but the records of those who know well what they are talking about ; neither are the writers in any sense pessimists. With all these warnings from men whose business it is to study the question and who are fully qualified to advance an opinion, surely it is time that we took up seriously the question of afforestation.

It may be said by some that the timber of our foreign

possessions will partly fill up the gap, but this is not the case. Indian timber, principally teak, is not in request to any appreciable extent, while the great African forests are hardwoods, and as a rule unsuited to our wants. The forests of South America are on a par with those of India and Africa, while China and Japan, as also Australia, require more timber than they possess.

Cost of suitable Land for Afforesting.—When in the past the question of afforesting has been brought forward, the usual outcry has been that suitable land is too expensive to buy. But this argument will no longer suffice, for, as I have before pointed out, excellent land for the cultivation of high-class timber can be procured in considerable quantity at about £2 per acre. Through the kindness of Lord Ancaster's estate agent, I have been allowed to look over the sale contracts of several parts of the Gwydyr Estate, in Carnarvonshire, and from these I find that 7,412 acres were disposed of, at an average price of £2 2*s*. 3*d*. per acre. The ground was excellent for the production of timber, as the larch on other adjoining lands clearly evidenced. Again the Crown recently purchased 12,500 acres in Scotland at the modest rental of about £2 per acre. Other instances could be quoted, but the above suffice to show that land in every way suited for profitable tree planting can be bought at probably less than £2 per acre.

It is perhaps unfortunate that many of these waste lands are private property, the owners of which, even if they could afford it, have little inclination to sink for a period of, say, twenty-five years the necessary capital required to be expended on the formation of plantations. But all this would be obviated by State ownership of the woodlands. Private individuals, or, indeed, public bodies, labour under many disadvantages in respect of afforestation, not the least, as before stated, being the quarter of a century required before the money expended in planting can be even partially recovered, while a systematic method of cultivation and large wooded areas are first necessities to successful timber culture. It is therefore preferable in every way that the Government should take up the question of tree planting on a large scale, the necessary land being available at a moderate cost per acre.

Cost of Forming Plantations.—This will vary greatly with the manner in which the work is carried out, the particular district of the country, nature of soil and rate of wages paid, as also whether fencing and draining have to be engaged in. The difference in cost between "notch" and "pit" planting is very considerable, and the fact that the former method is almost exclusively adopted on the rough grounds throughout Scotland accounts mainly for the smaller first outlay on Scottish plantations. Thus at Grantown, Strathspey, the Countess of Seafield's estate, Mr. Thomson, the very capable wood manager, has planted during the past forty-seven years upwards of 20,000 acres of woodlands, at a cost, including fencing, of rather under £2 per acre. In England, however, where, for various reasons, pit planting is adopted, and larger plants are used, the cost varies from £5 to £6 per acre. For all practical purposes, however, the cost of forming plantations may be put down at, say, £5 per acre, as an average taken from the following figures will show :—

England and Wales.

		£	s.	d.
Yorkshire, at 600 ft. altitude, cost of planting and fencing	per acre	4	18	9
Kent, fencing and planting .	,,	6	3	0
Lincolnshire ,, ,, .	,,	8	0	0
Gloucestershire ,, ,, .	,,	7	10	0
Carnarvonshire ,, ,, .	,,	5	2	0

Scotland.

		£	s.	d.
Inverness-shire, Glengloy Estate, 800 ft. altitude, cost of fencing and planting .	per acre	3	10	0
Ross-shire, up to 1,200 ft. altitude, cost of fencing and planting . . .	per acre	2	10	0
Perthshire (planting only) . .	,,	2	10	0
Blair Athol, 3,665 acres, fencing and planting	,,	2	10	0
Grantown, Strathspey, fencing and planting	,,	2	0	0

IRELAND.		£	s.	d.
Wicklow, 700–900 ft. altitude, fencing and planting	per acre	4	13	11
Armagh (bogland), fencing and planting	,,	5	2	0

Another instance in Scotland may be recorded, in which 550 acres were planted at a cost of £1,178, or at the rate of £2 2*s*. 10*d*. per acre. This included for fencing, £164 18*s*. 4*d*.; drainage, £123 15*s*.; plants, £520 10*s*.; planting, £368 16*s*. 8*d*.

In connexion with these figures, it may be reassuring to state that in each case a strict account of the expenditure involved had been carefully noted, and the returns given are practically correct. The average cost, therefore, taking Great Britain as a whole, would be about £5 per acre for fencing and planting the ground. The above-named plantations, too, were formed on the very class of ground of which we have so much lying idle or bringing in only a few shillings rental per acre, in various parts of the country. The Ross-shire plantation referred to was a bleak and barren moorland which the crofters, who used it as a common for their cattle and sheep, refused to rent at 1*s*. per acre per annum, while at Strathspey the 20,000 acres of land were let out previous to planting at 8*d*. per acre per annum. Vast tracts of the bare hillsides of Wales are only bringing in a few shillings of rental per acre. It should be remembered that all the above-named plantations were formed on bleak, exposed moorlands—the very class of waste lands that I have so strongly advocated as being suitable for the woodlands of the future, and of which at the present time there are about 15,000,000 acres lying idle in various parts of the kingdom. Therefore the cost of planting may be considered as or about £5 per acre. This, with £2 5*s*. for cost of purchase and 5*s*. for incidental expenses, would bring the initial total expenditure to £7 10*s*. per acre. Elsewhere I have suggested that 1,000,000 acres should be planted over a period of twenty-five years, at the rate of 40,000 acres per year, which would entail an outlay of £300,000 annually—a small sum when compared with the £25,000,000 expended each year by this country on supplies brought from abroad.

But there is another point that I should like to touch upon whilst dealing with the formation of plantations, and that is that the work should only be entrusted to the efficient and practical wood manager, who is fully conversant with the whole routine of woodland work. It is frequently urged that forestry does not pay, but where this holds good, the cause is always traceable to injudicious planting and wrong methods of management. No more can we expect the gardener, gamekeeper, estate joiner, or even the land agent to undertake economical timber culture than we could expect the forester to carry out successfully the duties of any of these individuals. Wrongly formed plantations are, unfortunately, far too common, in so far, at least, as adaptation of soil and trees are concerned, the result being that, financially speaking, the woods are a failure, and proprietors, in consequence, fight shy of further planting operations. When pressing home the question of the extension of plantations, I have more than once been confronted by the statement that past experience does not warrant further expenditure in that way. That this is true cannot be denied, in many instances at least, but, then, as above stated, faulty methods of management are alone responsible for the failure.

Financial Returns from Tree Planting.—Though it must be admitted that, in the majority of cases at least, the financial returns cannot be accepted as strictly correct (in most cases they are too low), owing to the woods being treated for other than commercial purposes, yet in not a few instances, where neither game rearing nor ornamental effect have to be considered, the yield of timber and gross returns for a stated number of years are perfectly reliable. Of course, where game coverts and underwood, or where the perfect development of the trees, as in ornamental plantations, are matters of first importance, and require that the individual specimens be scattered thinly over the ground, the greatest yield of the best quality of timber cannot be expected; but where, as on various Scottish and English estates, the trees are grown thickly together and solely for their economic value, the case is quite different, as the returns given below will attest.

One hundred acres of common land were planted from 1852 to 1862. Larch was the principal crop, with a few beech, Scotch pine, spruce and silver fir. The plantation was thinned at intervals from 1871 to 1884, the thinnings being sold for close on £500, but many trees were used for fencing and estate purposes generally. The whole plantation was felled in 1907, and realized fully £4,500, or at the rate of £45 per acre. The larch on the lower portion averaged 23¼ ft. per tree, but on the exposed ground the trees were only about one-third of that dimension. This plantation has a northern aspect, and is situated at from 800 ft. to 1,300 ft. above sea-level. After allowing for the cost of planting and interest on the money expended, the annual return per acre comes to about 20*s.* The adjoining heath-covered lands let for about 2*s.* 6*d.* per acre. Again, on the Countess of Seafield's estates, Scotland, on grazing land which formerly brought in 8*d.* per acre, Mr. Thomson, the woods manager, tells me that, at the age of forty-seven years, Scotch fir realized £40 per acre; while in another wood the individual trees brought 24*s.* 6*d.* each.

A larch plantation of 208 acres, on a steep hillside, was felled at the age of fifty years. The actual returns during that period were: from thinnings, £4,500; from final felling, £14,500; or fully £90 per acre. The original cost of planting was under £5 per acre, and the value of the land at thirty years' purchase £7 10*s.* per acre, thus leaving a balance of fully £78 per acre at the age of fifty years.

The extensive hillside plantations formed by the late Lord Powerscourt in Ireland, those at Glendalough in the same country, formed by the Duke of Atholl between Dunkeld and Blair Atholl, those at Glengoy, in Aberdeenshire, at Strathkyle in Ross-shire, and at Gwydyr and Penrhyn Castle in the Principality of Wales—all of which were formed over thirty-five years ago, account of the cost of formation and management being strictly kept—these surely afford sufficient evidence not only of the profitable returns to be obtained from woodlands, but of the feasibility of afforesting mountain lands with vast benefit in the way of shelter to the dreary, treeless, and bleak, exposed uplands where the planting has been carried out. As far as actual profits are concerned, it will be prudent to assume that for the first twenty years no return whatever will be derived from hillside plantations, the sales of thinnings up to that time barely covering the expense of cutting and interest on first cost. From twenty-five to forty years an annual return of fully 12*s*. per acre has in many instances been forthcoming, while the value of the standing crop at the latter age has been found to vary from £50 to £70 per acre. I do not think that these figures would be, generally speaking, too high, as at Balfour, in Scotland, the larch at forty-three years' growth on a hillside were valued at 20*s*. each, while a valuation of 21*s*. per tree was made of larch on the slopes of the Snowdon range of hills, in Wales, at the age of forty years. But many similar instances could be recorded, and are constantly coming before those who have to do with the valuing and felling of timber.

The late Mr. Lewis Miller, who had probably a larger experience of home woods than any other person, has given me some valuable and interesting information regarding what he has paid per acre for larch in various parts of Scotland. In twenty years, between 1870 and 1890, Mr. Miller has cut down growing timber to the value of over £250,000. A great many of the plantations were fifty years old, and yielded over £50 per acre when finally cut down, apart from the value of the thinnings taken out of them previously to the time they were cut down. To one proprietor in Aberdeenshire he paid £60,000 for plantations about fifty years of age, and the price worked out on an average at fully £50 per acre. One particular plantation of larch in Aberdeenshire, about seventy years old, yielded £150 per acre; another plantation, all larch, about forty-four years of age gave over £100 per acre, and these plantations were for the most part growing on what was formerly pasture or waste land, and cost for planting and fencing from £2 to £2 10*s*. per acre. It will be needless to multiply cases in which poor lands worth only from 1*s*. to 3*s*. per acre have been made to realize by judicious tree planting as much as 20*s*. per acre for fifty or sixty years with a final crop worth from £50 to

£75 per acre. All the plantations above referred to are excellent object-lessons of the possibilities of the British Isles for the production of high-class timber if woods are properly planted and managed.

Advantages of Tree Planting.—Not only from a strictly financial point of view but also from a hygienic sense standpoint, plantations are of the utmost importance.

For shelter for farm stock, for improving the agricultural value of the lower lying lands, and for the part they play in clothing and ornamenting our bare commons and hillsides, their value can hardly be over-estimated. Twenty-five years ago I formed a plantation on a spur of the Snowdon range of hills, in Wales, where the fierce, long-continued and hard-hitting blasts were of almost constant occurrence, and the amount of shelter and warmth it now affords to the farm stock and lower lying lands would hardly be credited. Previous to the formation of this particular plantation, at altitudes varying up to 600 ft., the adjoining lands were quite incapable of cultivation, but now crops are gradually creeping up the hillsides, while the farm stock find the much-needed shelter and warmth that they were formerly denied. So great has been the benefit of this wood both to man and beast that the farmer on whose land it was planted speaks of it as "a Godsend." Other similar cases in Wales might be mentioned, as for instance the plantations on the Gwydyr and Penrhyn Estates, and also those near Abergele, where the judicious planting up of rocky and almost worthless land has converted dreary and inhospitable districts into the most fashionable and expensive residential property. In many parts of Scotland, particularly Perth, Inverness and Aberdeenshire, equally good results have been obtained by judicious tree-planting.

Another notable advantage gained by the planting of trees lies in the provision of profitable work for the unemployed. As a special chapter is devoted to this important subject, it need only be mentioned here. It should not be overlooked that excellent results have followed in the wake of planting bog lands in Ireland. In 1862 my father formed several plantations there, a full account of which will be found in the *Transactions of the Highland and Agricultural Society of Scotland* for 1873. I examined these woods and was agreeably surprised at the height which the trees had attained, the cubic contents of the timber, and the price realized. Incidentally, it might be noticed that the ground previously to planting was a dreary, heath-clad waste, only suitable for snipe-shooting and the production of turf for fuel. Many other instances of the numerous advantages to be derived from a well-organized system of tree-planting could be cited, not the least important being the greater facilities that would be afforded for disposing of the timber. In many outlying districts all over the country far removed from road and rail, it is difficult to get rid of the small amount of timber that is periodically cut down, but were larger quantities handled and a continuity of supply forthcoming, I feel certain that timber merchants would be prompted to make special transit arrangements. More than once I have been asked by Irish landowners to recommend buyers of good larch and oak timber, but, after negotiation, I have invariably been told by the merchant that the quantity offered was far too small to allow of special facilities for delivery being provided, since the timber was far away from road and rail. They stated, however, that if a specified number of cubic feet of good timber could be guaranteed annually for a number of years they were quite prepared to buy. The same obstacles to the sale of timber have been experienced in Scotland and remote parts of Wales. In these cases a continuity of supply, such as would be forthcoming if my scheme of afforesting was carried out, would ensure speedy sales at moderate prices in places where at present it is difficult, if not impossible, to dispose of small quantities except at ruinously low prices.

INDEX

Q

R

S

Butler & Tanner Frome and London